AI FOR BEGINNERS

EFFORTLESSLY LEARN AI ESSENTIALS AND MASTER CUTTING-EDGE TRENDS IN JUST 10 DAYS TO FUTURE-PROOF YOUR SKILLS

ANNE FRANCES CARAANG

TABLE OF CONTENTS

Introduction   7

1. UNDERSTANDING AI BASICS   11
   1.1 Demystifying AI: From Sci-Fi to Reality   12
   1.2 How AI Mimics Human Intelligence   14
   1.3 Machine Learning Explained with Everyday
   Examples   16
   1.4 Deep Learning in a Nutshell: Teaching
   Machines to Learn   18
   1.5 Neural Networks: The Backbone of AI   20
   1.6 The Role of Data in AI: Fueling the Future   23

2. AI IN OUR DAILY LIVES   27
   2.1 How AI Curates Your Social Media Feeds   27
   2.2 AI Behind the Scenes: Online Shopping
   Recommendations   30
   2.3 Smart Assistants: How AI Powers Siri and
   Alexa   33
   2.4 AI in Healthcare: From Diagnosis to Treatment   35
   2.5 The Magic of AI in Navigation Apps   38
   2.6 AI and Your Privacy: What You Need to Know   41

3. THE BUILDING BLOCKS OF AI   45
   3.1 Algorithms Unveiled: The Decision-Makers
   of AI   46
   3.2 Understanding Data Mining and Pattern Relief   49
   3.3 The Significance of Big Data in AI
   Development   51
   3.4 Breaking Down AI Bias and How It Affects Us   54
   3.5 Ethics in AI: Navigating the Moral Landscape   56
   3.6 Security and AI: Protecting Against Cyber
   Threats   59

4. HANDS-ON AI PROJECTS FOR BEGINNERS   63
   4.1 Creating Your First AI Chatbot   63
   4.2 Introduction to DIY AI Projects at Home   66

4.3 Beginner's Guide to AI in Web Development   68

4.4 AI and Photography: Enhancing Images
with AI   71

4.5 Building a Simple AI to Play Your Favorite
Games   73

4.6 AI in Content Creation: Writing Articles
with AI   75

5. AI TOOLS AND TECHNOLOGIES   79

5.1 Exploring AI with Google's AI Tools   80

5.2 Leveraging OpenAI's Platforms for Beginners   83

5.3 AI Development Kits for Non-Coders   85

5.4 Simple AI Models You Can Train at Home   88

5.5 Using Cloud Services for AI Projects   91

5.6 Understanding and Using AI APIs for Projects   95

6. AI FOR CAREER ADVANCEMENT   99

6.1 AI Skills That Employers Are Looking For   99

6.2 How to Showcase Your AI Projects on Your
Resume   101

6.3 Networking in the AI Community: Tips and
Tricks   103

6.4 Transitioning Your Career to AI and Tech   106

6.5 Continuous Learning: Keeping Up with AI
Trends   108

6.6 Freelancing with AI Skills: Finding Your Niche   111

7. EMERGING TRENDS AND FUTURE
PREDICTIONS   115

7.1 Generative AI: The Next Frontier   115

7.2 The Role of AI in Sustainable Technologies   118

7.3 AI in Space Exploration: What Lies Ahead   120

7.4 The Future of AI in Education   123

7.5 Ethical AI: Shaping a Responsible Future   125

7.6 The Convergence of AI and Virtual Reality   128

8. BUILDING AN AI-INCLUSIVE WORLD   131

8.1 AI for Social Good: Case Studies and
Opportunities   132

8.2 Bridging the Digital Divide with AI Education   135

8.3 The Global Impact of AI on Employment   137

8.4 Advocating for Diversity in AI Development   139

8.5 AI Accessibility: Making Tech Usable for
Everyone   141
8.6 Envisioning an AI-Empowered Society: Our
Role and Responsibility   143

Conclusion   147
References   151

INTRODUCTION

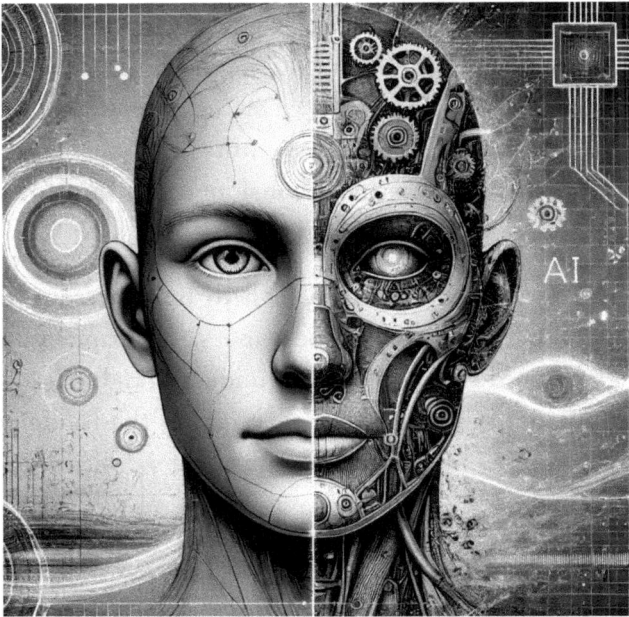

Did you know that whenever you use a streaming service to binge your favorite show, there's a sophisticated algorithm curating content specifically for you? This is just one of the many ways artificial intelligence (AI) seamlessly integrates into our lives, often without us even realizing the depth of its influence.

My journey into the world of AI began with curiosity mixed with a healthy dose of skepticism. Like many of you, I was intimidated by the jargon and complexity that seemed to gate-keep the true power of AI technologies. However, as I navigated through these initial hurdles, my apprehension became a profound passion for making AI understandable and accessible to everyone. I firmly believe that understanding AI is not just for those with technical backgrounds but is crucial for everyone navigating this digital age.

In this book, "AI for Beginners: Effortlessly Learn AI Essentials and Master Cutting-Edge Trends in Just 10 Days to Future-Proof Your Skills," I aim to peel back the layers of complexity surrounding artificial intelligence. This guide is crafted to transform intricate AI concepts into clear, engaging, and actionable insights. We will explore not just the nuts and bolts of AI but also its practical applications, ethical considerations, and its transformative potential in our daily lives.

You don't need a background in tech to benefit from this book. I've designed it especially for beginners, ensuring all technical terms are broken down into simple, understandable language. By the end of this journey, you'll not only grasp the fundamentals of AI but also appreciate its implications and applications in your everyday life.

Let me share a personal story that underscores the potential of AI. I remember the first time I encountered AI through ChatGPT. It was an exciting moment filled with curiosity and a touch of apprehension. I've always been fascinated by technology and how it could mimic human thought processes. When I sat down to have a conversation with ChatGPT, I didn't know what to expect.

At first, it felt like chatting with a knowledgeable friend who had a seemingly endless reservoir of facts and stories. I asked questions about history, science, and even quirky topics like the best way to care for a bonsai tree. The responses were not only accurate but also surprisingly engaging. It was clear that this AI could understand and generate language in a way that felt almost human.

But the most memorable part of that experience was when I decided to test its creativity. I asked ChatGPT to help me brainstorm ideas for a short story. The AI's suggestions were not just imaginative; they were thought-provoking and rich with potential. It felt like a collaborative effort, as if I was bouncing ideas off a creative partner.

That first experience with ChatGPT was a revelation. It opened my eyes to the incredible potential of AI in enhancing creativity, learning, and everyday problem-solving. It was the beginning of a journey into understanding the capabilities and possibilities of artificial intelligence, and it left me eager to explore more.

This book stands out because it doesn't just focus on the technicalities. We will consider how AI shapes our society, the ethical lines it treads, and the future it is steering us toward. Expect hands-on projects, insights into real-world applications, and narratives on how today's AI innovations are being utilized across different sectors.

I invite you to approach this book with an open mind and a keen sense of curiosity. Consider it your first step towards demystifying AI, a tool that is not only reshaping our present but will also define our future. Let's embark on this enlightening journey together, ready to unlock the vast potential that AI holds.

As we move forward, I am excited about the possibilities that learning about AI can bring to your life and to our world. AI is not just a field for engineers and tech experts—it's a burgeoning landscape ready for all of us to explore, influence, and innovate. Together, let's discover how AI can help solve some of our most pressing challenges and usher in a future brimming with opportunity.

CHAPTER ONE

UNDERSTANDING AI BASICS

As you step into the realm of artificial intelligence, you might find yourself surrounded by a whirlwind of terminologies and concepts that seem straight out of a science fiction novel. It's fascinating, isn't it? How concepts once limited to imaginative storytelling have now permeated the very fabric of our daily lives, influencing everything from the way we shop to how we communicate. In this chapter, we'll unravel the layers of AI, turning what might seem like complex science into understandable and relatable information. You'll discover the roots of AI, address common misunderstandings, and appreciate the profound impact AI has on society. By the end of this chapter, the once daunting world of artificial intelligence will begin to feel a bit more familiar and significantly less intimidating.

1.1 DEMYSTIFYING AI: FROM SCI-FI TO REALITY

The Evolution of AI: Tracing AI's Journey from a Concept in Science Fiction to its Current Status as a Pivotal Technology

Artificial Intelligence, or AI, began as a mere figment of the collective imagination of pioneering science fiction writers. Early literary works depicted sentient machines with capabilities far beyond human intelligence, sparking curiosity and fear alike. Fast-forward to the mid-20th century, the term "Artificial Intelligence" was officially coined at a conference at Dartmouth College in 1956, marking the commencement of AI as a legitimate field of scientific research. The journey from then to now is filled with both spectacular successes and humbling setbacks. Today, AI is not just a theoretical study; it's a crucial component of the technology that powers our everyday lives, from smart assistants like Siri and Alexa to more profound applications such as disease prediction tools in healthcare.

Common Misconceptions about AI: Dispelling Myths and Clarifying What AI Can and Cannot Do

One prevalent myth is that AI can function independently of human intervention. On the contrary, AI systems are designed and operated by humans and can only operate within the parameters set by their creators. Another common misconception is that AI is a looming threat to human employment. While AI does automate some tasks, it also creates new jobs and industries, many of which require human oversight and strategic thinking. Understanding these nuances helps demystify AI and encourages a more nuanced perspective on its role in our world.

The Basic Principles of AI: Introducing the Core Ideas That Underlie All AI Technologies

At its core, AI is about creating systems that can perform tasks that would typically require human intelligence. These tasks include reasoning, interpretation, and decision-making. AI operates on a foundation of algorithms and data - the algorithms allow machines to carry out specific tasks, and the data provides the information needed to make decisions. Machine learning, a subset of AI, takes this a step further by enabling systems to learn and improve from experience without being explicitly programmed.

AI's Impact on the Society: Exploring How AI Has Begun to Transform Industries, Economies, and Daily Life

AI's influence is vast and varied. In healthcare, AI algorithms help diagnose diseases with high accuracy and speed. In the automotive industry, AI drives the development of autonomous vehicles. In entertainment, AI algorithms personalize content recommendations on platforms like Netflix and Spotify. Economically, AI is a powerhouse of efficiency, streamlining operations, reducing costs, and boosting productivity across multiple sectors. Socially, while AI brings numerous benefits, it also presents challenges such as privacy concerns and ethical dilemmas, which are critical to address as we integrate AI more deeply into our societal structures.

This exploration of AI's evolution, its foundational principles, common myths, and societal impact aims to provide you with a clear and balanced understanding of artificial intelligence. As we navigate through this chapter, remember that AI, at its essence, is a tool created by humans to enhance and enrich our interactions with the world.

1.2 HOW AI MIMICS HUMAN INTELLIGENCE

Understanding cognitive AI begins with appreciating how it attempts to replicate human thought processes. Consider how we, as humans, learn and adapt from the experiences we encounter; our brains process inputs, make connections, and derive conclusions, which in turn guide our future actions. Cognitive AI aims to mimic this ability, but through a technological framework. By utilizing algorithms that analyze data and learn from patterns, AI systems can perform tasks such as recognizing speech, interpreting complex data, and making informed decisions. For instance, when you interact with a digital voice assistant, the AI behind it processes your words, comprehends the query, and fetches the information or performs the action you requested. This is a simplistic form of how AI emulates human-like processing, bridging the gap between human cognitive abilities and machine execution.

Moving to the comparison between machine learning and human learning, the contrasts and similarities offer profound insights. Humans learn through experience, education, and by using our senses, which is a dynamic and often non-linear process. We are capable of abstract thinking and can learn from very few examples or even from hypothetical scenarios. On the other hand, machine learning, particularly in its most common form—supervised learning—requires large amounts of data to learn. These AI systems learn from examples rather than through direct programming. They adjust their responses based on the accuracy of their predictions, which improves over time as they process more data. However, unlike humans, AI typically requires a vast quantity of specific examples to learn effectively and can struggle with abstract concepts or anything that deviates significantly from the data it was trained on.

The role of algorithms in AI cannot be overstated—they are fundamental to enabling machines to make decisions. Algorithms are sets of rules or instructions that AI systems follow to perform tasks and solve problems. Think of them as the recipe that guides the system on how to analyze data and make decisions. For example, recommendation algorithms on streaming services analyze your viewing history and use that data to suggest shows you might like. These algorithms continually refine their predictions based on your interactions, becoming more accurate over time. The efficiency, reliability, and scalability of AI systems heavily depend on the robustness of their underlying algorithms, which are crafted to handle specific tasks and scenarios.

Despite the remarkable capabilities of AI, it's important to acknowledge the limitations of AI in mimicking human intelligence. Current AI technologies excel in handling specific tasks that can be defined by clear rules and have access to abundant data. However, AI lacks general intelligence, meaning it does not possess understanding or consciousness. AI systems cannot replicate the emotional depth or ethical reasoning humans bring to decision-making processes. They are also limited by the data they are trained on, which can lead to biases or errors if the data is not comprehensive or is skewed. Moreover, AI struggles with tasks that require common sense, intuition, or creative thinking, areas where human intelligence naturally excels.

As we continue to integrate AI into various facets of life and industry, recognizing these boundaries is crucial. It helps in setting realistic expectations about what AI can achieve and ensures that we remain mindful of the areas where human oversight is irreplaceable. By advancing our understanding of both the capabilities and limitations of AI, we can better leverage this powerful technology to complement human intelligence, rather than attempting to replace it.

1.3 MACHINE LEARNING EXPLAINED WITH EVERYDAY EXAMPLES

Machine learning, a term that often surfaces in discussions about AI, might sound complex, but its concepts are part of everyday technology, influencing many services you use regularly. At its core, machine learning is a method of data analysis that automates analytical model building. It's a branch of artificial intelligence based on the idea that systems can learn from data, identify patterns, and make decisions with minimal human intervention. To put it simply, machine learning enables computers to perform specific tasks from the data they analyze, without being explicitly programmed to perform those tasks.

Let's break down the two main types of machine learning: supervised and unsupervised learning, which are distinguished by their approaches to teaching a computer model how to make predictions or decisions. Supervised learning occurs when we teach the machine using data that is well-labeled, meaning the data is tagged with the correct answer. It's akin to a teacher guiding a student through a textbook with the answers at the back. For instance, in spam detection, a supervised learning algorithm analyzes thousands of emails that are pre-labeled as 'spam' or 'not spam' and learns to predict the classification of new emails. Unsupervised learning, on the other hand, deals with data that has no historical labels. The system is not told the "right answer." The algorithm must figure out what is being shown. The goal is to explore the data and find some structure within. It's like a student in a library discovering patterns or topics without a teacher's guidance. A common example is customer segmentation in marketing, where algorithms divide customers into groups with similar characteristics without prior annotation, purely based on underlying patterns in the data.

Machine learning powers many familiar services, integrating seamlessly into your daily activities, often without your explicit awareness. Take, for example, Netflix's recommendation system. This service uses supervised learning to predict the rating you might give a show based on your viewing history and similar decisions made by other users. The system analyzes patterns in your behavior and continuously adjusts the recommendations to better suit your preferences. Similarly, email services use machine learning to filter out spam, learning from various indicators such as message headers or the frequency of certain words to predict whether an email is spam.

Looking ahead, the future of machine learning is incredibly promising and is poised to revolutionize many aspects of our lives. As machine learning technologies continue to evolve, they will become more efficient at processing and analyzing vast amounts of data. This will enhance their ability to make more accurate predictions and decisions in real-time, which could redefine capabilities in everything from medical diagnostics to personal finance management. Moreover, as machine learning algorithms become better at unsupervised learning, they will unlock new potentials in understanding complex patterns without needing large labeled datasets. This could significantly reduce the barriers to AI implementation in fields where data labeling is costly or impractical.

Machine learning is not just a futuristic concept; it's a present reality that is shaping your interaction with the digital world. Whether it's by curating the content you watch, helping filter out unnecessary emails, or even guiding autonomous vehicles, machine learning stands as a fundamental pillar of modern AI applications. As we continue to innovate and integrate these technologies, the ways in which they learn from and adapt to our needs will only become more refined, making machine learning an even more integral part of our technological landscape.

1.4 DEEP LEARNING IN A NUTSHELL: TEACHING MACHINES TO LEARN

Deep learning, a term that has gained tremendous traction in the tech community and beyond, is often perceived as a complex and elusive concept. At its essence, deep learning is an advanced subset of machine learning. It uses layered (hence 'deep') neural networks to analyze various factors of data. These layers, composed of nodes, mimic the human brain's structure and function, allowing machines to recognize patterns and characteristics in data similar to the way humans do. What sets deep and machine learning apart is the ability of deep learning models to handle and interpret vast amounts of unstructured data such as images, sound, and text, making it a powerful tool for many of today's AI innovations.

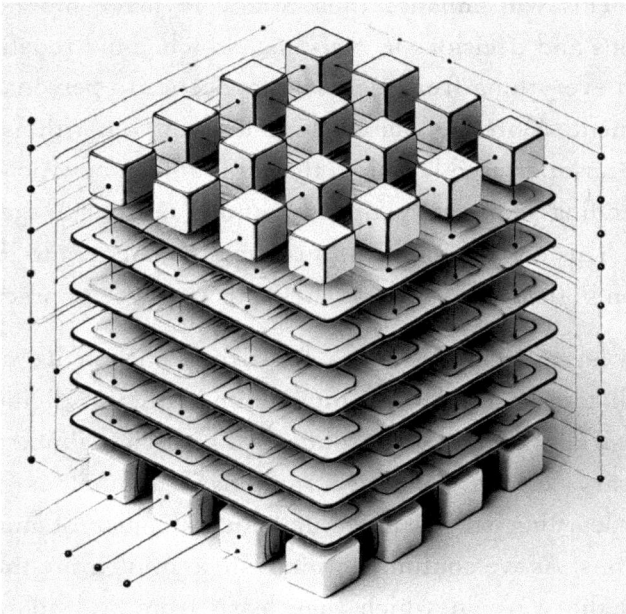

To understand the architecture of neural networks, imagine a vast network of neurons in the human brain. Each neuron in a neural network processes a small bit of task-specific information. This information is then passed along to other neurons in the network. The connections between these neurons can strengthen or weaken over time, influenced by a process known as 'learning.' This learning occurs through exposure to vast amounts of data, and over time, the network adjusts its internal parameters to make better predictions. In practice, this means a deep learning model designed for image recognition will gradually learn to recognize and differentiate between elements in pictures more accurately, refining its understanding as it processes more images.

Applications of deep learning are both profound and wide-ranging, touching industries from healthcare to entertainment. In image recognition, deep learning models power the facial recognition systems used in various applications from unlocking smartphones to enhancing security systems. In the realm of voice assistants like Siri and Alexa, deep or machine learning algorithms help parse human speech with remarkable accuracy, enabling these tools to understand and respond to user commands. Moreover, deep learning is instrumental in the development of autonomous vehicles. It helps cars perceive their surroundings, make decisions, and navigate roads with little or no human intervention, using a process known as computer vision which is heavily reliant on deep learning.

However, the journey of deep learning is not without its challenges. One significant limitation is the requirement for extensive computational resources. Training deep learning models often requires substantial amounts of data and considerable computing power, which can be cost-prohibitive. There's also the challenge of 'black box' algorithms, where the decision-making process within deep learning models is opaque and not easily understood by humans. This can lead to trust issues, especially in critical applications like medical diagnostics or in scenarios requiring moral and ethical decision-making. Despite these challenges, the opportunities presented by deep learning continue to spur developments in the field. Researchers are continuously exploring more efficient ways to train these models, reduce their dependency on vast datasets, and make their workings more transparent.

The potential future developments in deep learning are particularly exciting. Imagine a scenario where medical professionals could predict diseases even before symptoms appear, thanks to predictive models developed through deep learning. Or consider the possibility of truly personalized education, where learning materials are automatically adapted to suit the unique needs of each student, powered by deep learning algorithms that understand and interpret each student's interactions and progress. These advancements could significantly alter our approach to some of the most pressing challenges in healthcare, education, and beyond.

As deep learning continues to evolve, its integration into daily technology and operations seems set to increase, holding the promise of more intuitive interfaces, smarter applications, and machines that understand and interact with the world in increasingly sophisticated ways. As we harness these technologies, our interaction with machines will likely become more fluid and natural, mirroring human intuition closer than ever before.

1.5 NEURAL NETWORKS: THE BACKBONE OF AI

Neural networks, a fundamental concept in AI, draw their inspiration directly from the vast network of neurons in the human brain. Just as neurons in the brain activate in response to stimuli, transmitting signals to one another and creating pathways that inform responses and actions, neural networks in AI systems connect layers of artificial neurons that process input data, learn from it, and produce output. This biomimicry extends beyond mere structural imitation; it encapsulates the very essence of learning and adaptation that characterizes human cognition. Each 'neuron' in an artificial neural network receives input, processes it through a mathematical function, and passes the output to the next layer. The process is dynamically adjusted as the network 'learns' from each input it receives, enhancing its accuracy over time without explicit instructions on how to improve. This ability to adapt and learn from data makes neural networks particularly powerful in handling complex tasks such as speech recognition, where the nuances of language and human communication patterns can be vast and varied.

The diversity of neural network architectures is vast, each tailored to specific types of tasks and data. The most common type is the feedforward neural network, where connections between the nodes do not form cycles. This type is extensively used for straightforward prediction tasks, where inputs are directly connected to an output. Another significant architecture is the recurrent neural network (RNN), which is better suited for sequential data like time series analysis or natural language processing. In RNNs, connections between nodes form a directed graph along a temporal sequence, allowing them to exhibit temporal dynamic behavior. Unlike feedforward neural networks, RNNs can use their internal state (memory) to process sequences of inputs. This makes them ideal for tasks like language modeling and text generation, where the sequence and context of previous information are crucial. Convolutional Neural Networks (CNNs), another important architecture, are primarily used in processing pixel data, and are therefore prevalent in image and video recognition tasks. They employ a mathematical operation called convolution which enables them to focus on specific aspects of the input data, making them highly efficient for tasks involving image classification, face recognition, and even video analysis.

Training neural networks involves adjusting the weights of connections between the neurons to minimize the difference between the predicted output and the actual output, a process referred to as 'learning'. This training process typically requires a large amount of data, which the network uses to gradually improve its predictions or classifications. The data is divided into batches, and for each batch, the network makes predictions, checks the accuracy of these predictions, and adjusts the weights accordingly. This is often done using a method called backpropagation, where the error is propagated back through the network, providing insight into how each neuron's weights need to be adjusted to reduce errors. The goal is to refine these weights to a point where the network can not only replicate training data but can also generalize from it to make accurate predictions on new, unseen data. This ability to generalize is the true test of a neural network's efficacy and is crucial in practical applications where conditions can vary significantly from the training data.

The impact of neural networks on AI development cannot be overstated. They have profoundly transformed the capabilities of AI systems, enabling previously unimaginable advances. In healthcare, neural networks are used to diagnose diseases with greater accuracy than ever before, analyzing medical images, and identifying patterns that might escape human notice. In the automotive industry, they are crucial to the development of autonomous driving technologies, where they process vast amounts of sensory data in real time to make driving decisions. In the realm of customer service, chatbots powered by neural networks can understand and respond to a wide range of human inquiries, providing a level of interaction that closely mimics human conversation. The versatility and adaptability of neural networks have thus been central to the current wave of AI innovations, pushing the boundaries of what machines can learn and achieve. Their continued development and application are likely to drive significant transformations across various sectors, enhancing efficiencies and creating new opportunities for innovation and growth.

1.6 THE ROLE OF DATA IN AI: FUELING THE FUTURE

If we consider artificial intelligence as an intricate machine, data is undoubtedly the fuel that powers it. Without data, AI systems cannot learn, adapt, or make decisions. To understand why data is indispensable for AI, imagine teaching someone to recognize various fruits without ever showing them any examples. Without data—these examples—it would be impossible for them to learn. Similarly, AI systems require vast amounts of data to learn how to carry out tasks. This data can come from numerous sources: images, text, clicks on a website, medical records, and more. Each piece of data helps the AI system to learn about patterns and nuances, which is crucial for tasks such as translating languages, recommending products, or diagnosing diseases.

However, the process of collecting and managing this data comes with its own set of challenges and responsibilities. Collecting vast amounts of data often involves sophisticated technology and methodologies. Organizations must handle this data with care, ensuring accuracy and relevance. For example, a company developing an AI system for facial recognition needs diverse data that represents different ethnicities, ages, and lighting conditions to perform well across various scenarios. Managing this data also involves ensuring it is stored securely and organized in a way that is accessible for AI systems while maintaining user privacy. Best practices in data collection and management are critical, not just for the effectiveness of AI systems, but also for maintaining public trust and compliance with regulations.

One of the significant risks in data-driven AI development is the introduction of bias. Bias in AI occurs when an algorithm produces systematically prejudiced results due to erroneous assumptions in the machine learning process. For instance, if an AI model for credit scoring is trained primarily on historical data from a neighborhood that has experienced financial instability, it may unjustly penalize residents of that neighborhood. This kind of bias can perpetuate and even exacerbate social inequalities. Mitigating this risk requires a conscientious effort in designing, collecting, and processing data. Techniques such as using diverse datasets, regularly testing AI systems for bias, and developing AI with transparency in mind are essential steps towards reducing bias.

The ethical considerations surrounding AI data usage are vast and complex. As AI systems become more prevalent in our lives, concerns about how data is used, who has access to it, and how it affects privacy are increasingly critical. Ethical AI use starts with transparency about how data is collected and used, and ensuring that it is done with the consent of those involved. Moreover, there should be a clear benefit to the users, and they should have control over their data, including the ability to correct or remove it. Privacy must be a priority, not an afterthought, in the design and implementation of AI systems. As developers and users of AI, we are on the front lines of navigating these ethical challenges. It is our responsibility to advocate for and implement practices that protect individual rights and promote trust in AI technologies.

In the unfolding landscape of AI, data remains a foundational element that will continue to shape the capabilities and impact of AI systems. As we advance, our approach to handling data will significantly determine the trajectory of AI's integration into society. Responsible data practices are not just beneficial; they are imperative for the sustainable and ethical growth of AI technologies. By prioritizing these practices, we can harness the full potential of AI to innovate and improve lives while safeguarding individual rights and societal values.

CHAPTER TWO

AI IN OUR DAILY LIVES

I magine this: You wake up in the morning, reach for your phone, and while scrolling through your social media feed, every post seems to be just what you were interested in. How does your social media know you so well? The answer lies not in some mystical psychic ability, but in the sophisticated use of artificial intelligence that curates content just for you. This chapter will explore how AI integrates into everyday activities, influencing not only what you see online but also enhancing your interactions with the digital world while carefully balancing privacy considerations.

2.1 HOW AI CURATES YOUR SOCIAL MEDIA FEEDS

Personalization Algorithms: Unveiling How AI Tailors Content to Individual Preferences

Every time you like a post, watch a video, or even pause on a specific content piece, AI algorithms are working in the background, learning your preferences to create a tailor-made social media experience just for you. These algorithms are a type of machine learning technology designed to analyze patterns in your behavior. They adjust what you see on your feed based on what they predict will engage you the most. For instance, if you frequently watch cooking videos, the algorithm notes this and subsequently, you might find more recipes and culinary content flooding your feed. This is AI's way of ensuring that the content you see matches your interests, thereby enhancing your engagement with the platform.

The Mechanics of Content Recommendation: Exploring How AI Decides What Shows Up in Your Feed

The process of content recommendation involves complex algorithms that track your interactions and compare them with the data from millions of other users. This process, known as collaborative filtering, helps the AI system to identify and predict content that you might enjoy, based on similarities with others who have a viewing history like yours. For example, if you and several other users follow the same fitness influencer, the algorithm might suggest other popular health and wellness pages that those users also engage with, assuming you share similar interests. These sophisticated AI systems can analyze vast amounts of data in real-time, continuously learning and adapting to provide the most relevant content to each user.

The Impact on User Experience: Assessing How AI-Curation Affects Our Interaction with Social Media

The impact of AI-driven personalization on user experience is profound. By delivering content that is tailored to individual preferences, social media platforms can increase user engagement significantly. This not only makes the platform more enjoyable and relevant for you but also benefits advertisers by placing products in front of the eyes most likely to be interested in them. However, there is a delicate balance to strike. Over-personalization can lead to the creation of a "filter bubble," where you are only exposed to ideas and viewpoints that conform to your existing beliefs. This can limit exposure to diverse perspectives and potentially reinforce misconceptions or biases.

Privacy Considerations: Discussing the Balance Between Personalized Content and User Privacy

While the benefits of personalized content are clear, it raises significant privacy concerns. The vast amount of data collected by AI systems includes detailed information about your personal preferences, behaviors, and even location. This data can be incredibly valuable, not just to advertisers, but potentially to malicious actors if not properly secured. Therefore, transparency about what data is collected, how it is used, and who it is shared with is crucial. Many social media platforms provide settings that allow you to control the level of personalization and data sharing, but the onus is also on you to understand and utilize these tools. Ensuring a balance between personalized experiences and privacy protection remains a critical challenge in the design and regulation of AI systems in social media.

In this exploration of AI's role in curating social media content, it's clear that AI can significantly enhance user experience by personalizing content to match individual preferences. However, as we navigate this digital age, understanding and managing the trade-offs between personalization and privacy is essential. As AI continues to evolve, fostering a more profound understanding of these mechanisms and their implications can empower you to make more informed choices about how you interact with technology daily.

2.2 AI BEHIND THE SCENES: ONLINE SHOPPING RECOMMENDATIONS

When you're browsing through an online store and suddenly see a product that feels like it was specifically picked for you, there's a good chance AI is behind that moment of serendipity. AI-driven recommendation systems have become the backbone of the e-commerce industry, subtly guiding your shopping experience by suggesting products that you might find appealing. These systems use complex algorithms to predict your preferences and show you items that align with your past behavior, potential interests, and even search patterns.

Understanding these recommendation systems requires a grasp of how they work at a fundamental level. Essentially, when you interact with an online platform—be it searching, clicking, or purchasing—each action is tracked and analyzed. This data feeds into algorithms that process and learn from your behavior to build a model of your preferences. Over time, the system becomes adept at predicting what products might catch your eye next. For instance, if you've been looking at a lot of kitchenware, the AI might start showing you new and popular kitchen gadgets, assuming that these would be relevant to your interests. This method is not just about pushing sales; it's about creating a personalized shopping experience that makes browsing more enjoyable and efficient for you.

The role of your data in these recommendation engines is significant. Every click you make contributes to a profile that AI systems develop, which is used to tailor your shopping experience. This might raise questions about privacy, which are valid and important. However, in the context of enhancing service, the data allows for a more customized experience. If you frequently purchase books in a particular genre, the AI uses this data to highlight upcoming releases and deals within that genre, thereby saving you time and effort in finding these items yourself. It's a proactive approach to customer service, where the platform anticipates your needs and meets them without needing to be asked.

However, the integration of AI in e-commerce isn't without its challenges. While the convenience of having personalized recommendations can enhance your shopping experience, there's an inherent risk of over-personalization. Similar to the filter bubbles on social media, you might find yourself seeing a narrower selection of products, which can limit your exposure to different options that could potentially catch your interest. Moreover, if not properly managed, the data collected can be susceptible to breaches, leading to privacy concerns. Ensuring that these systems are secure and that data is handled responsibly is crucial for maintaining trust and integrity in AI-powered e-commerce.

The future of AI in shopping looks to be as dynamic as it is innovative. With advancements in machine learning and data analysis, the next generation of recommendation systems could offer even more nuanced insights into consumer behavior. Imagine a scenario where AI not only recommends products based on what you've liked in the past but also incorporates external factors such as current trends, local weather, and even economic forecasts to make predictions. This could transform how businesses manage inventory and develop new products, making them more responsive to market conditions and consumer needs.

Furthermore, as AI technology advances, there's potential for these systems to become more interactive. Instead of passively receiving recommendations, you could have a dialogue with an AI assistant that understands your preferences and assists you in making shopping decisions, much like a personal shopper would. This could make online shopping more engaging and personalized, providing an experience that rivals and perhaps surpasses in-store shopping.

The integration of AI into online shopping has already made significant impacts on both the consumer experience and business operations. As we look to the future, these technologies promise to further refine and enhance the way we shop online, making it more personalized, efficient, and enjoyable.

2.3 SMART ASSISTANTS: HOW AI POWERS SIRI AND ALEXA

In a world where convenience is king, smart assistants like Siri and Alexa have become central figures in many of our daily routines. But have you ever stopped to wonder how these devices can respond to a wide array of questions and commands? The secret lies in the sophisticated use of AI technologies, specifically voice recognition and natural language processing (NLP). These technologies allow your smart assistant to not only hear but understand and process your requests, making daily tasks more efficient.

Voice recognition technology is the first step in the interaction between you and your smart assistant. This technology works by converting your spoken words into text that the AI system can understand. The process begins with the device detecting a 'wake word'—like "Hey, Siri" or "Alexa"—which activates the assistant. Following activation, the smart assistant begins recording your speech, which it sends to a powerful AI server. Here, your speech is quickly transformed into text. This might seem straightforward, but the technology must contend with various accents, speech nuances, and background noises, all of which can affect the accuracy of the transcription. Continuous improvements in AI algorithms have significantly enhanced this technology, enabling more accurate and rapid recognition of diverse voices and commands.

Once your speech is transcribed, the real magic happens with Natural Language Processing, or NLP. This facet of AI allows machines to understand and interpret human language in a meaningful way. NLP involves several processes, such as parsing (analyzing the grammatical structure of sentences), semantic analysis (understanding the meanings of words), and context recognition (understanding the context in which words are used). For example, if you ask, "Will it rain today?" NLP helps the assistant understand that 'it' refers to the weather at your location on this specific day. The assistant can then respond appropriately based on weather data it pulls from the internet. This level of understanding is continually refined as NLP technologies learn from vast amounts of linguistic data, improving the assistant's ability to respond more accurately to increasingly complex queries over time.

The convenience offered by smart assistants is undeniable. They provide hands-free assistance with various tasks such as setting reminders, playing music, controlling smart home devices, and providing real-time information like news and weather updates. This functionality can significantly simplify daily activities, particularly for those with hectic schedules or mobility challenges. Imagine cooking dinner while simultaneously setting reminders for your week's appointments and adjusting your home's lighting system—all through voice commands. The integration of smart assistants into our homes and smartphones saves time and reduces the need to interact manually with multiple devices and applications.

However, the integration of AI-powered smart assistants in our lives isn't without concerns, particularly regarding security and privacy. These devices are essentially always listening for their wake word, which raises questions about what else they might inadvertently record. While manufacturers assure that these devices only record and process audio after hearing the wake word, accidental activations can occur, leading to potential privacy breaches. Moreover, the data collected through these interactions, such as personal preferences and daily routines, can be immensely valuable for companies in terms of targeted advertising and service customization. Ensuring that this data is securely stored and not misused is paramount. Users must be provided with clear information about how their data is used and must have control over their privacy settings, including the ability to review and delete their data history.

As AI technology evolves, the capabilities of smart assistants will continue to expand, further blurring the lines between human and machine interaction. The challenge for developers and users alike will be to navigate these advancements responsibly, ensuring that as these technologies become more embedded in our lives, they do so in a way that respects our privacy and enhances our daily experiences without compromising our security.

2.4 AI IN HEALTHCARE: FROM DIAGNOSIS TO TREATMENT

In the vast and intricate world of healthcare, AI is increasingly playing a pivotal role, particularly in the realms of diagnostics and treatment. The application of artificial intelligence in these areas is not just about technological advancement; it's about fundamentally transforming how medical care is delivered, making it faster, more accurate, and far more personalized than ever before. When it comes to diagnostics, AI systems are being trained to read and interpret everything from X-rays to genetic sequences at speeds and with a level of precision that humans cannot match. For instance, AI-driven tools are now used to analyze diagnostic images like CT scans and mammograms. These tools can detect subtle patterns that might be overlooked by the human eye, identifying signs of diseases such as cancer at earlier stages when treatment is more likely to be successful. This capability enhances the accuracy of diagnoses and significantly accelerates the process, enabling quicker decision-making that can be critical to patient outcomes.

The power of AI extends beyond diagnostics into the realm of personalized medicine, which promises treatments tailored specifically to individual patients. This personalization is made possible by AI's ability to analyze vast amounts of data, including a patient's own genetic makeup, lifestyle, and previous health records. By integrating and interpreting this data, AI systems can help predict how different patients will respond to various treatments, allowing healthcare providers to choose the most effective therapies for each individual. This approach is a shift from the one-size-fits-all strategy traditionally used in medicine to one that is finely tuned to each patient's unique biological profile, potentially increasing the effectiveness of treatments and minimizing side effects.

Moreover, AI is streamlining a myriad of administrative tasks in healthcare settings, significantly enhancing operational efficiencies. Hospitals and clinics are deploying AI systems to manage everything from patient appointments to medical records and billing. These systems automate routine tasks that previously required extensive human labor, freeing up healthcare professionals to focus more on patient care rather than paperwork. For example, AI-powered scheduling systems can predict peak times for patient visits and help clinics allocate their staff resources more efficiently, reducing wait times and improving patient satisfaction. Similarly, AI applications in medical coding and billing can reduce errors and streamline the claims process, making it faster and more accurate, which benefits both healthcare providers and patients.

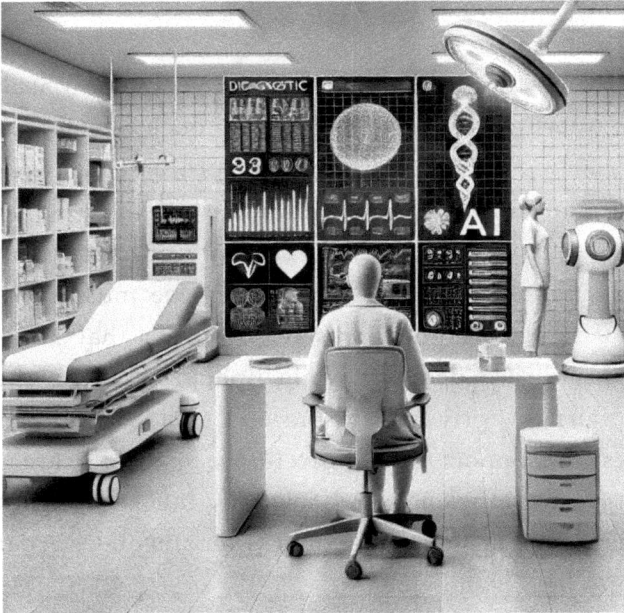

However, the integration of AI into healthcare also brings several ethical considerations that must be carefully managed. Ensuring fairness in AI-driven diagnostics and treatments is paramount, as these systems can inadvertently perpetuate biases present in the training data. For instance, if an AI system is trained primarily on data from one demographic group, it may be less accurate for people outside that group. Therefore, it's crucial to use diverse data sets that reflect the broad spectrum of patients seen in clinical settings. Privacy is another significant concern, as the use of AI in healthcare involves handling sensitive personal data. Protecting this data against breaches and ensuring it is used ethically is essential to maintain trust in AI applications. Moreover, there must be accountability in AI-driven decisions, especially in high-stakes areas like healthcare. It's important for patients and providers to understand how AI tools make their recommendations and to have mechanisms in place to review and challenge these decisions if necessary.

As AI continues to evolve, its potential to transform healthcare is immense, offering possibilities that were once the stuff of science fiction. From revolutionizing diagnostic processes to personalizing treatment plans and streamlining administrative workflows, AI is set to play an increasingly central role in healthcare. However, as we harness these advanced technologies, we must also navigate the ethical landscape they present, ensuring that AI serves to enhance healthcare delivery while upholding the highest standards of fairness, privacy, and accountability. Embracing these challenges and opportunities, the future of healthcare promises not only greater efficiency and accuracy but a new paradigm of personalized medicine that could fundamentally change our approach to health and wellness.

2.5 THE MAGIC OF AI IN NAVIGATION APPS

Navigating through bustling city streets or finding the quickest route to a new destination has become remarkably simpler, thanks to AI-driven navigation apps. These apps utilize sophisticated route optimization algorithms that analyze multiple factors to suggest the best routes. At the core of these algorithms lies a dynamic decision-making process that evaluates various parameters such as distance, traffic conditions, and road types. For instance, when you enter a destination, the AI assesses possible routes based on current traffic data, road closures, and even weather conditions to recommend the most efficient path. This isn't just about finding the shortest physical distance; it's about calculating the quickest or easiest route considering all real-time conditions, which can significantly differ from one minute to the next.

The real-time traffic assessment is another area where AI proves invaluable. By continuously gathering and analyzing data from various sources including satellite images, sensors on roads, and information from other users, AI systems provide up-to-date traffic information that can alter route recommendations on the fly. This capability is crucial during peak traffic hours or unexpected road incidents such as accidents or construction. For example, if an accident occurs on a route you're currently taking, the AI can immediately analyze the situation and reroute you to avoid delays, thereby saving you time and reducing road congestion. This dynamic response enhances individual convenience and contributes to better overall traffic management in urban areas.

Moreover, the integration of AI in navigation tools goes beyond mere convenience; it significantly enhances road safety. By providing real-time updates and alerts about road conditions, speed limits, and potential hazards, AI-powered apps help drivers make informed decisions that can prevent accidents. They can alert drivers to slow down in bad weather conditions or when approaching heavy traffic, helping maintain a safe driving environment. Additionally, for those who are navigationally challenged, voice-guided directions and alerts mean that drivers can keep their eyes on the road without needing to look at a map, reducing the chances of distraction-related incidents.

Looking ahead, the future of autonomous driving heavily relies on the advancements in AI navigation technologies. Self-driving vehicles, which once seemed like a futuristic fantasy, are becoming a reality, with AI at their helm. These vehicles use AI not only for navigating roads but also for detecting and responding to their surroundings. Through a combination of sensors, cameras, and advanced AI algorithms, autonomous cars can perceive their environment, make split-second decisions, and navigate safely. The ongoing development of these AI systems focuses on improving their ability to understand complex scenarios, such as interpreting the gestures of a traffic officer or navigating through chaotic urban environments without human intervention.

As these technologies continue to evolve, the potential for fully autonomous vehicles becomes more tangible. AI's ability to learn from vast amounts of data and its capability to make real-time decisions play a crucial role in this evolution. Future advancements may allow AI to handle even more complex driving tasks, further pushing the boundaries of what autonomous vehicles can do. This has implications for individual users and promises significant changes in the logistics and transportation industries, where efficiency and safety are paramount. As we move forward, the integration of AI in navigation and driving technologies is set to redefine our approach to travel, commuting, and transport, making it safer, more efficient, and less stressful.

2.6 AI AND YOUR PRIVACY: WHAT YOU NEED TO KNOW

In today's digital age, the intersection of artificial intelligence (AI) and privacy is a crucial topic for anyone interacting with modern technology. While AI brings undeniable benefits, enhancing convenience and efficiency across various aspects of our lives, it also raises significant privacy concerns that cannot be overlooked. The core of the issue lies in the trade-off between enjoying personalized, AI-driven experiences and maintaining control over our personal information. As AI systems require vast amounts of data to function optimally, from personal preferences to behavioral patterns, the potential for privacy breaches increases. Understanding this trade-off is essential for navigating the AI-integrated world responsibly.

The balance between convenience and privacy is a delicate one. AI's ability to streamline our daily activities, provide personalized recommendations, and even predict our needs is grounded in its capacity to analyze personal data. Every interaction you have with AI-enhanced devices or services feeds into a cycle of data collection and analysis, enabling these systems to become better at predicting and meeting your needs. However, this same data can also be a vulnerability. For instance, if not properly managed and secured, the information collected can be accessed by unauthorized parties or even misused, leading to potential privacy violations.

Recognizing these risks, significant efforts have been made to implement robust data security measures. These include encryption, which secures data by transforming it into a code that can only be deciphered with a specific key, and data anonymization, which involves stripping personally identifiable information from the data sets, ensuring that the individuals' privacy is maintained even if the data is intercepted. Additionally, AI systems are increasingly designed with built-in privacy features, such as automatic data purging policies that delete information that is no longer necessary. These measures are crucial for protecting personal information but are only as effective as their implementation and the continuous updates they receive to tackle evolving cyber threats.

Moreover, the regulation of AI and data privacy has become a focal point for governments worldwide. Laws and regulations such as the General Data Protection Regulation (GDPR) in the European Union and the California Consumer Privacy Act (CCPA) in the United States have set precedents for how personal data should be handled and protected. These regulations grant individuals greater control over their data, offering rights such as the ability to access, correct, and even request the deletion of their data. They also impose strict penalties on organizations that fail to protect consumer data, incentivizing better data practices. However, the landscape of data privacy laws is complex and varies significantly across different regions, which can be challenging for both individuals and businesses to navigate.

In addition to relying on security measures and regulations, there are practical steps you can take to safeguard your privacy in an AI-driven world. Firstly, being informed about the privacy policies of the services you use can help you understand how your data is being collected, used, and protected. Opting to adjust the privacy settings on your devices and applications to limit data sharing is another proactive measure. Furthermore, consider regularly reviewing the permissions granted to applications on your devices, revoking access that is not necessary for the functionality you use. Lastly, staying informed about the latest in data security and AI developments can empower you to make better decisions about your data and privacy.

As AI continues to evolve and integrate further into the fabric of daily life, the conversation around privacy must not only continue but also deepen. Balancing the benefits of AI with the need to protect personal privacy is an ongoing challenge, requiring not just technological solutions but also a strong ethical framework and informed public discourse. By understanding the trade-offs, taking proactive steps to protect personal information, and participating in broader conversations about AI and privacy, individuals can better navigate the complexities of this digital age.

As we close this chapter and move forward, the insights shared here lay the groundwork for a more profound understanding of AI's role in society and the ethical considerations that come with its advancements. From enhancing user experiences to the potential risks involved, the dual-edged nature of AI technology presents a landscape ripe for exploration and continuous learning. As you step into the following chapters, keep these nuances in mind, and consider how AI's broader applications might influence not just personal privacy but the societal fabric as a whole.

CHAPTER THREE

THE BUILDING BLOCKS OF AI

I magine walking into a grand orchestra where each musician's performance is meticulously orchestrated by a conductor. Now, replace those musicians with various AI technologies, and the conductor with something far less visible yet profoundly influential: algorithms. In the vast concert that is artificial intelligence, algorithms are the unsung maestros, directing the flow of data and decision-making processes that enable AI systems to function. This chapter peels back the layers of these critical components, offering you a clear view of what algorithms are, how they evolve, and the pivotal role they play in the operatic symphony of AI applications that touch our lives.

3.1 ALGORITHMS UNVEILED: THE DECISION-MAKERS OF AI

The Definition and Importance of Algorithms: Clarifying What Algorithms Are and Why They're Central to AI

At its core, an algorithm is a set of instructions or rules designed to perform a specific task or solve a problem. Think of it as a recipe in a cookbook, where each step guides you through the process of creating a dish. In the realm of AI, algorithms are these recipes that guide computers on how to process data and make decisions. For instance, when you use a search engine, algorithms work behind the scenes to determine which results are displayed based on the keywords you've entered. These algorithms are fundamental because they provide the step-by-step instructions that AI systems need to function effectively. Without them, AI would be like a kitchen filled with ingredients but no recipes to follow.

How Algorithms Evolve: The Process of Training and Refining AI Algorithms

The evolution of AI algorithms is a dynamic process that involves continuous training and refinement to improve their performance. This training process is akin to how a chef might refine a recipe over time, adjusting ingredients and techniques until the desired outcome is achieved. In AI, this involves feeding algorithms large sets of data, on which they can practice and learn. Through a method known as machine learning, these algorithms analyze the data, understand patterns, and gradually improve their ability to make predictions or decisions. For example, an AI algorithm used for facial recognition will become more accurate as it processes more images of faces. This learning process allows AI to adapt and become more efficient in tasks ranging from automated customer service to complex medical diagnostics.

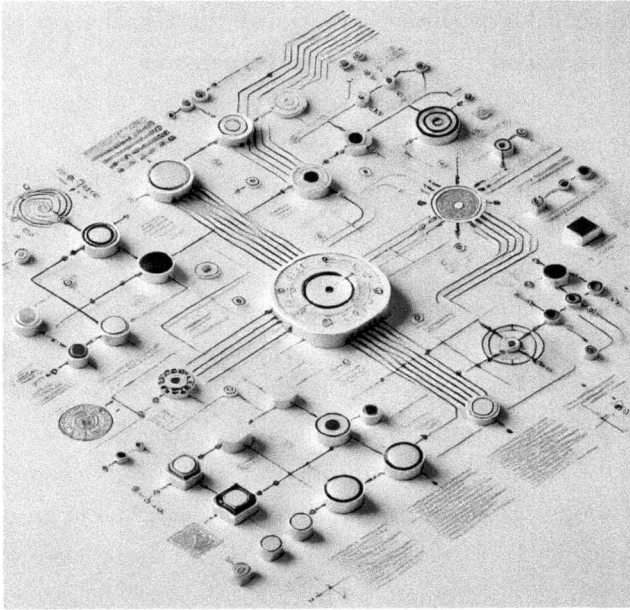

Examples of AI Algorithms in Action: Illustrating How Algorithms Power Search Engines, Social Media, and More

To see algorithms in action, look no further than the search engine you use every day. When you query a search engine, algorithms instantly sift through billions of web pages to find the most relevant results based on your search terms. These algorithms rank these results based on several factors, including relevance to your query, page quality, and user engagement, to ensure that you find what you are looking for. In social media, algorithms determine what content appears in your feed. By analyzing your interactions —likes, shares, time spent on posts—these algorithms curate a personalized feed designed to keep you engaged on the platform. Each interaction you make informs the algorithm, helping it to learn more about your preferences and refine the content it shows you.

The Limits of Algorithms: Discussing the Challenges and Limitations Faced by Current AI Algorithms

Despite their capabilities, AI algorithms have their limitations. One significant challenge is their dependence on data. The quality and quantity of data used to train algorithms greatly impact their effectiveness. Poor data or data that is not representative of the real world can lead to inaccurate or biased results. For instance, if a facial recognition algorithm is trained mostly on images of people from a single ethnic group, it may not perform well on faces from other ethnicities. Additionally, algorithms are typically designed for specific tasks and may not perform well outside their intended scope. This specialization means that while an algorithm can be excellent at recommending movies based on your past viewing habits, it cannot be repurposed to diagnose medical conditions without substantial modifications and additional training.

As we continue to explore the building blocks of AI, understanding algorithms—their design, function, and limitations—is crucial. These sets of rules are what make it possible for AI systems to perform tasks that range from mundane to extraordinarily complex. By demystifying these core components, we not only gain insight into how AI works but also appreciate the intricacies of developing and managing these powerful tools. As we delve deeper into the components that constitute AI, remember that each element, much like the musicians in an orchestra, plays a critical role in the symphony of interactions that drive modern technology.

3.2 UNDERSTANDING DATA MINING AND PATTERN RELIEF

Data mining is a process through which large sets of data are analyzed to uncover hidden patterns, unknown correlations, and other useful information. It's akin to a treasure hunt, where the treasure is not gold or jewels, but insights that can lead to better decision-making in business, science, medicine, and virtually any other field. The significance of data mining in AI cannot be overstated; it is the groundwork upon which smarter systems are built, allowing machines to uncover patterns and insights without human help, thereby enhancing their learning and predictive capabilities.

Pattern recognition forms the core of many AI operations and can be mostly seen as the ability of machines to identify and categorize data based on either previous examples or through the development of self-learned patterns. This capability is what enables your email to filter out spam, your bank to detect fraudulent transactions, or health professionals to identify tumors in medical imaging. The process starts with the collection of data, which is then analyzed and broken down into various features. These features are then used to develop models that can recognize patterns and make predictions. For example, in image recognition, an AI system breaks down the image into pixels and identifies patterns in the pixel arrangement to determine what the image represents.

The applications of data mining and pattern recognition stretch across various industries, revolutionizing the way we handle data and make decisions. In finance, these technologies are used for risk management, fraud detection, and customer relationship management. Banks and financial institutions mine data to understand and predict customer behavior, assess creditworthiness, and detect anomalous transactions that could indicate fraud. In healthcare, data mining helps in diagnosing diseases, predicting hospital readmissions, and developing personalized treatment plans. By analyzing patterns in patient data, healthcare providers can identify risk factors for diseases and tailor treatments that specifically match the patient's profile, thereby improving outcomes.

However, the power of data mining and pattern recognition is not without its concerns. The potential for misuse of these technologies is significant. Data mining can invade privacy, leading to the unauthorized use of personal information. For instance, companies might mine personal data to target users more aggressively or even manipulate behavior, raising ethical concerns. Moreover, if not properly managed, these technologies can perpetuate biases present in the data they are trained on. This can lead to unfair outcomes, such as discriminatory practices in hiring, lending, and law enforcement. Therefore, ethical data mining practices are crucial. These involve ensuring transparency in the algorithms used for data analysis, obtaining data through fair and legal means, and continually checking and correcting for biases.

The importance of integrating ethical considerations into data mining practices is becoming increasingly recognized. Organizations are now more than ever required to not only focus on how effectively they can mine data but also on how ethically they do it. This involves setting up guidelines that govern data mining operations, ensuring the privacy and security of data, and maintaining fairness in automated decisions. In this rapidly evolving field, staying ahead means not only leveraging the capabilities of data mining and pattern recognition to enhance operational efficiency and decision-making but also doing so in a manner that respects privacy and promotes fairness. As we continue to explore the vast capabilities of AI, understanding and implementing ethical data mining will be crucial in harnessing its full potential responsibly and sustainably.

3.3 THE SIGNIFICANCE OF BIG DATA IN AI DEVELOPMENT

In the vast landscape of AI development, big data is not just a buzzword but a fundamental pillar that significantly enhances the capabilities of artificial intelligence systems. Let's demystify what big data is—it refers to massive datasets that may be analyzed computationally to reveal patterns, trends, and associations, especially relating to human behavior and interactions. This data is culled from myriad sources like internet transactions, GPS signals, social media interactions, and more, each adding layers of complexity and richness to the data pool. The role of big data in AI is transformative; it provides the raw material from which AI systems can learn and make informed decisions. For example, in healthcare, big data derived from patient health records, treatment plans, and recovery outcomes offers invaluable insights that AI uses to predict patient risks and personalize healthcare treatments.

Handling such voluminous and diverse data sets is not without its challenges. The sheer volume of big data means that traditional data processing applications are inadequate. Specialized tools and technologies are required to store, manage, and analyze this data effectively. Moreover, big data is often unstructured and messy. It comes in varied formats – from structured, numeric data in traditional databases to unstructured text documents, emails, videos, audios, stock ticker data, and financial transactions. Organizing this unstructured data in a way that makes it usable for AI systems is a significant challenge, requiring sophisticated algorithms and computing power. The velocity of big data also adds to the complexity; data streams in at unprecedented speeds and must be dealt with in a timely manner. For instance, data from social media streams requires real-time processing to gain insights into consumer behavior and market trends, necessitating highly responsive AI tools that can operate at the speed of information.

The accuracy of AI predictions and decisions is directly proportional to the quality and quantity of data it learns from. Big data offers a more comprehensive view of a problem by providing a larger set of examples for AI to learn from, which can lead to more accurate and robust AI systems. For instance, consider a voice recognition system. The more voice samples it processes, the better it becomes at understanding accents, dialects, and languages, enhancing its accuracy. Similarly, an AI system for fraud detection trained on a vast array of transaction data can more effectively identify patterns that indicate fraudulent activities. However, while big data can enhance the accuracy of AI systems, it also raises issues of data quality. Poor quality data can lead to incorrect conclusions, making it imperative that data is accurately processed and cleansed before it is used for training AI systems.

Looking towards the future, the symbiotic relationship between big data and AI is set to deepen. The ongoing evolution in big data analytics promises to bring even more sophisticated tools to the fore, enabling finer, more nuanced insights. AI, in turn, will become more adept at handling complex, multi-layered data sets, opening up new possibilities for innovation across various sectors. We can anticipate developments such as advanced predictive analytics, where AI can forecast future trends with high accuracy based on historical big data, or augmented decision-making, where AI assists humans in making complex decisions by providing detailed data-driven insights. This evolving landscape suggests a future where big data and AI not only coexist but collaboratively push the boundaries of what technology can achieve in understanding and serving human needs more profoundly. As we continue to explore the vast potentials of AI, the role of big data as a crucial enabler of more intelligent, responsive, and efficient AI systems becomes ever more apparent, promising a future where both data and AI converge to create transformative impacts across all walks of life.

3.4 BREAKING DOWN AI BIAS AND HOW IT AFFECTS US

In the intricate weave of artificial intelligence development, one thread that requires meticulous attention is the issue of AI bias. This bias, often unintentional, can be embedded in AI systems, affecting their decisions and operations. Understanding where these biases originate and how they perpetuate is crucial for developing fair and effective AI systems. Essentially, biases in AI can originate from various sources, primarily from the data used to train these systems. Since AI learns from data, any inherent prejudices in the data will likely be learned and amplified by the AI. For instance, if a dataset used to train a recruitment AI is composed predominantly of profiles from a particular gender or ethnic group, the AI may develop a skewed understanding of what a 'suitable' candidate looks like, leading to biased hiring decisions.

The perpetuation of bias in AI can also stem from the design of the algorithms themselves. If the creators of these algorithms have conscious or unconscious biases, these can inadvertently be coded into the AI systems. Another significant factor is the feedback loop where biased AI decisions generate more data, which then feeds back into the system, reinforcing these biases. For example, if an AI in a social media platform is more likely to recommend certain types of content to specific demographics, it might limit the diversity of content those users are exposed to, reinforcing narrow content loops.

The consequences of such biased AI systems are profound and far-reaching. On an individual level, biased AI can lead to unfair treatment in various sectors including job recruitment, loan approvals, and legal sentencing. In healthcare, AI bias can lead to misdiagnoses or inappropriate treatment plans if the data on which the AI was trained is not representative of all demographic groups. These biases can erode trust in AI technologies, limiting their potential benefits and leading to societal disparities. On a broader scale, the perpetuation of biases through AI can exacerbate existing social inequalities, embedding discrimination deeper within societal structures. This affects the groups directly discriminated against and undermines the social fabric by fostering divisions and injustice.

Mitigating AI bias is therefore not just a technical necessity but a moral imperative. Strategies to reduce bias in AI systems include diversifying the data used in training these systems. Ensuring that the data reflects a wide range of demographics can help in developing AI systems that are fair and equitable. Another effective strategy is the implementation of regular audits for AI systems. These audits can be conducted by independent third parties who assess AI decisions for fairness and accuracy, ensuring that any biases are identified and corrected promptly. Additionally, involving diverse teams in the development of AI algorithms can provide multiple perspectives that help in identifying potential biases that might not be evident to a more homogenous group.

Highlighting the role of diversity in combating AI bias underscores an essential approach to AI development. Diversity in AI development teams is not only about gender, race, or ethnicity but also about varied educational backgrounds, life experiences, and professional expertise. This variety can lead to more comprehensive deliberations and considerations in the design and implementation of AI systems, ensuring that they serve the needs of a diverse population effectively. For instance, when developing an AI system for global markets, having team members from different countries can provide insights into local cultural contexts that might affect the functionality and reception of the AI. This can prevent cultural biases and ensure that the AI system is adaptable and sensitive to global diversity.

As we delve deeper into the nuances of AI development, understanding and addressing AI bias is critical for harnessing the technology's full potential. By acknowledging the sources of bias and implementing strategies to mitigate these, the development of AI can be steered towards more equitable and just outcomes. This not only enhances the reliability and acceptability of AI systems but also ensures that they contribute positively to societal advancement. As AI continues to evolve and integrate into various aspects of life, it is imperative that these technologies are guided by principles of fairness and inclusivity, ensuring they benefit all of society equitably.

3.5 ETHICS IN AI: NAVIGATING THE MORAL LANDSCAPE

The rise of artificial intelligence brings not only technological advancement but also complex ethical dilemmas that impact society in profound ways. As AI systems become more integral to our daily lives, the moral implications of their decisions and actions come under scrutiny. One of the pressing ethical questions is the extent to which AI should have autonomy in decision-making, especially in critical areas such as healthcare, law enforcement, and military applications. For instance, should an autonomous vehicle make decisions about potentially life-threatening situations? The moral complexities of such scenarios highlight the need for a robust ethical framework in AI development.

Creating ethical AI involves establishing principles and guidelines that ensure AI technologies are developed and used in ways that benefit society while minimizing harm. This process begins with transparency, ensuring that AI operations are understandable by the users and that the decisions made by AI systems can be explained. Accountability is another crucial principle, where stakeholders, including developers and companies, are responsible for the outcomes of AI systems. This accountability is vital for building trust and ensuring that AI technologies are used responsibly. Moreover, fairness must be a cornerstone in AI development, ensuring that AI systems do not perpetuate or exacerbate social inequalities. Ensuring fairness involves rigorous testing across diverse scenarios to identify and mitigate any form of bias that the system might learn from the data it is fed.

Real-life case studies where AI ethics were brought into question provide tangible insights into the challenges and complexities of ethical AI. One notable example is the use of AI in recruitment processes, where several companies have utilized algorithmic systems to screen job applicants. However, instances have surfaced where these systems inadvertently discriminated against certain groups of people, such as older candidates or those from specific ethnic backgrounds, due to biases in the training data. These cases serve as crucial lessons on the importance of ethical considerations in AI deployment, highlighting the need for continuous monitoring and evaluation to ensure that AI systems perform their tasks without unfair biases or discrimination.

The future of AI governance is another critical area that requires attention as AI technologies increasingly permeate various sectors of society. The need for regulation and oversight is paramount to ensure that AI development aligns with societal values and ethical standards. This governance involves not only national and international laws but also industry standards that regulate the deployment of AI. The challenge lies in crafting regulations that are flexible enough to encourage innovation, while stringent enough to prevent misuse of the technology and protect public welfare. As AI continues to evolve, ongoing dialogue among technologists, policymakers, legal experts, and the public will be essential to navigate the ethical landscape and guide the responsible development and use of AI.

In navigating these ethical waters, it is crucial for everyone involved—from AI developers to end-users—to be informed and engaged in the ethical dimensions of AI. By fostering an inclusive dialog that considers diverse perspectives and values, we can hope to steer AI development in a direction that respects human dignity and promotes a just and equitable society. As we continue to explore and expand the capabilities of artificial intelligence, let us also commit to a path of ethical vigilance and responsibility, ensuring that our technological advancements reflect our deepest values and aspirations.

3.6 SECURITY AND AI: PROTECTING AGAINST CYBER THREATS

In this digital age, the role of AI in cybersecurity has become akin to that of a vigilant sentinel, guarding against a spectrum of cyber threats that evolve as rapidly as the technology itself. AI's ability to analyze vast quantities of data at incredible speeds is invaluable in identifying and responding to potential security threats in real time. For instance, AI systems are employed to monitor network traffic for unusual patterns that might indicate a cyberattack, such as an unusually high amount of data being transferred out of the network, which could suggest a data breach. Furthermore, AI-driven security protocols are capable of learning from each incident, continuously improving their detection algorithms to recognize new and evolving threats. This adaptive capability is crucial given the dynamic nature of cyber risks, where new malware or hacking strategies are constantly developed.

However, the integration of AI into cybersecurity is not without its risks. AI systems themselves can become targets for cyberattacks. One prominent risk is the poisoning of AI models through manipulated data inputs that lead to flawed learning outcomes. If a hacker can influence what the AI system learns, it can potentially control its behavior, turning a protective measure into a vulnerability. For example, an AI system designed to filter emails for phishing attempts might be tricked into categorizing malicious emails as safe if it is trained on tampered data. This vulnerability underscores the importance of securing AI systems and ensuring that the data they learn from is well-guarded and from reliable sources.

Balancing the innovation in AI with security is a delicate yet crucial endeavor. As AI technologies advance, ensuring these systems are secure must be a priority, not an afterthought. This involves adopting a security-by-design approach, where security measures are integrated into the development process of AI systems from the outset, rather than being tacked on as a final step. It also means investing in robust encryption technologies that protect data inputs and outputs from tampering, and conducting thorough security audits regularly to identify and address vulnerabilities before they can be exploited.

Public awareness about the role of AI in cybersecurity and the potential risks involved plays a crucial role in enhancing both personal and organizational security. Individuals and organizations must be informed about the ways AI systems can protect against cyber threats, as well as the measures necessary to secure AI operations. Education on recognizing potential AI vulnerabilities and understanding the basics of AI security can empower users to ask the right questions and make informed decisions about the AI technologies they adopt and rely on. This informed engagement is essential as it fosters a culture of security and encourages a proactive approach to cyber defense, which is vital in a landscape where cyber threats are ever-present and ever-evolving.

In summary, as we navigate through the complexities of AI and cybersecurity, understanding the dual role of AI as both a protector against and a target for cyber threats is crucial. By integrating robust security measures into AI systems and fostering public awareness, we can leverage AI's potential to enhance cybersecurity while safeguarding these technologies from being compromised. As we close this chapter on the foundational elements of AI, we've unraveled the intricacies of algorithms, data handling, ethical considerations, and now, the pivotal role of security in AI. These components collectively shape the development and application of AI technologies, influencing their effectiveness, reliability, and impact on society. As we move forward, the insights gained here lay a solid groundwork for exploring the transformative applications of AI across various sectors, which we will delve into in the upcoming chapters. This journey into AI's capabilities and its profound implications continues to unfold, promising to redefine the boundaries of what technology can achieve.

CHAPTER FOUR

HANDS-ON AI PROJECTS FOR BEGINNERS

D iving into artificial intelligence can be exhilarating yet daunting, especially when you start to think about creating your own AI projects. But what if you could build a virtual assistant or a chatbot, not in some high-tech lab, but right from the comfort of your home, using tools that are both accessible and user-friendly? In this chapter, we will embark on an exciting venture into the world of AI chatbots. These digital creations can converse, assist, and provide company, acting as a first step into the broader universe of AI for many beginners.

4.1 CREATING YOUR FIRST AI CHATBOT

Introduction to Chatbot Development: Simplifying the Process of Building a Basic AI Chatbot

The idea of creating an AI chatbot might sound complex, but it's quite accessible once you break down the process. A chatbot is essentially a software application used to conduct an on-line chat conversation via text or text-to-speech, in lieu of providing direct contact with a live human agent. Designed to simulate the way a human would behave as a conversational partner, chatbot systems typically require a vast amount of input data to learn from. However, for a basic chatbot, you can use predefined input patterns and responses. This means you can create a functional chatbot by defining a set of rules and responses without needing to dive deep into more complex AI technologies like machine learning from the get-go.

Choosing the Right Platform: Evaluating Different Platforms and Tools Available for Chatbot Creation

When building your first chatbot, selecting the right platform is crucial. There are numerous platforms available today that cater to non-experts, offering drag-and-drop interfaces to create chatbots without writing a single line of code. Platforms like Chatfuel,

Botsify, or ManyChat provide user-friendly interfaces and various customization options, making them ideal for beginners. These platforms also offer integration capabilities with popular messaging applications such as Facebook Messenger, WhatsApp, or Slack, allowing your chatbot to reach a wider audience effortlessly.

Designing a Conversation Flow: Tips for Designing Engaging and Useful Conversation Flows for Your Chatbot

The heart of a chatbot is its conversation flow—the exchanges it has with users. To design an effective conversation flow, start by defining the purpose of your chatbot. Whether it's answering frequently asked questions, booking appointments, or providing entertainment, the purpose will guide the dialogue structure. Start with a greeting and basic interactions to engage users and make them comfortable. Use a friendly tone and keep the flow logical and straightforward. Tools like decision trees or flowcharts can be helpful in mapping out the conversation paths before implementing them in the chatbot platform.

Testing and Improving Your Chatbot: Strategies for Testing Your Chatbot and Refining Its Responses Based on User Feedback

Once your chatbot is set up, testing becomes essential. Begin by testing it yourself, and then invite friends or family to try it. Observe how the chatbot handles different queries and whether the conversation flows smoothly. User feedback is invaluable, so consider integrating a feedback mechanism within the chatbot to collect users' opinions. Based on the feedback, you can refine the chatbot's responses, add new features, or tweak the conversation paths. Continuous testing and iteration are key to enhancing the chatbot's effectiveness and user experience.

Incorporating these elements into your first AI chatbot project will not only provide a practical introduction to AI but also a fulfilling experience as you watch your digital creation interact and evolve. This hands-on project serves as a foundation for more complex AI endeavors, offering a glimpse into the mechanics of automated interactions and the satisfaction of bringing a virtual conversationalist to life. As you progress, remember that every adjustment and improvement is a step forward in your AI development journey, bringing you closer to mastering the art of artificial intelligence in everyday applications.

4.2 INTRODUCTION TO DIY AI PROJECTS AT HOME

Embarking on a do-it-yourself AI project at home can be an exhilarating way to apply your newfound knowledge in a practical, hands-on manner. The first step, selecting your first project, is crucial and should be approached with your interests and the scope of practical application in mind. For beginners, it's wise to start with something not only captivating but also manageable. A project that balances simplicity and the opportunity to expand your skills is ideal. Think about what excites you about AI — whether it's the ability to automate tasks, analyze data, or something else — and use that as a starting point for choosing your project. For instance, if you're fascinated by how AI can automate everyday tasks, you might consider a project that uses AI to sort your emails or manage your calendar.

Once you have an idea, the next step is gathering the necessary resources and tools. This phase is like assembling your paint and brushes before you start a painting. For most basic AI projects, you'll need access to software platforms that can facilitate AI development without heavy coding. Platforms such as Google's AIY Projects offer kits and online support to help beginners implement AI into simple projects using pre-written code and user-friendly instructions. Additionally, open-source libraries like TensorFlow or Scikit-learn provide tools to more adventurous beginners ready to dabble in a bit of coding. These libraries come with comprehensive documentation and active community forums that can help troubleshoot issues or provide inspiration.

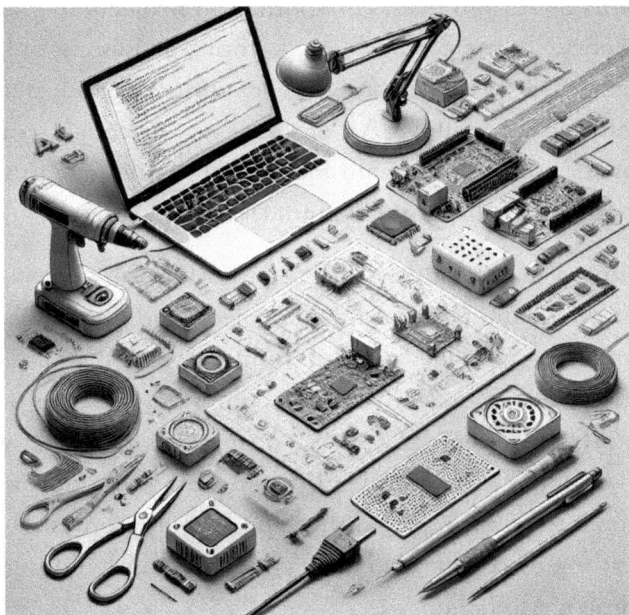

To get your creative juices flowing, here are a few project ideas that are well-suited for beginners: building a simple recommendation system, creating a basic voice command device, or developing a rudimentary image recognition system. Each of these projects allows you to explore different aspects of AI, from natural language processing to computer vision, and they can be scaled according to your comfort level and interest. For example, a simple recommendation system can be started by creating a program that suggests books based on the user's past reading habits. This project can later be expanded to include more complex variables like user ratings, genres, or author styles.

Engaging with the community is an invaluable part of learning, especially when diving into something as expansive as AI. Online forums, social media groups, and local meet-ups can be goldmines of information, inspiration, and support. Platforms like GitHub offer access to a plethora of projects where you can see real-world applications of AI and even collaborate on projects to gain hands-on experience. Participating in hackathons or joining online challenges can also provide practical experience and networking opportunities with fellow AI enthusiasts. Through community engagement, you not only gain insights and guidance but also the encouragement to push your boundaries and innovate.

By selecting a manageable project, gathering the necessary tools, starting with a simple but scalable project, and engaging with the AI community, you lay a solid foundation for your DIY AI journey at home. These initial steps help demystify the process of applying AI and open up a world of creative possibilities where your newfound skills can lead to meaningful and innovative applications. As you progress, these projects not only bolster your understanding of AI but also your confidence in handling increasingly complex AI challenges.

4.3 BEGINNER'S GUIDE TO AI IN WEB DEVELOPMENT

Artificial intelligence has seamlessly woven itself into the fabric of web development, transforming it into an arena where static pages evolve into interactive, intelligent experiences that engage users on a whole new level. If you're stepping into web development or looking to infuse existing projects with a touch of AI, understanding how AI can enhance web applications is crucial.

AI in web development focuses on improving user experience, automating processes, and personalizing interactions, making websites not just informational but truly adaptive and responsive to user needs.

Introducing AI into your web projects starts with identifying the aspects of your website that can benefit most from automation or enhanced intelligence. For instance, AI can be used to personalize user experiences by analyzing user behavior and providing content recommendations similar to those seen on streaming services. This personalization makes each user's interaction with the website unique and more engaging, likely increasing retention rates and user satisfaction. Similarly, AI can be implemented to improve search functionalities on your website. By using natural language processing, the search feature can become more intuitive, understanding user queries conversationally and providing more accurate results. This capability can be particularly beneficial for e-commerce sites, where enhanced search functionality can directly influence customer satisfaction and sales.

For those eager to dive into implementing AI in web projects, the process can be straightforward with the right tools and a clear plan. Begin by defining the scope of AI implementation: decide what functionalities you want to enhance with AI and what outcomes you expect. Once the scope is defined, the next step is integrating AI functionalities. This often involves selecting and setting up AI-powered tools and APIs that align with your project's needs. For instance, if you're looking to add a chatbot to your website, platforms like Dialogflow or Microsoft Bot Framework provide powerful tools to create and deploy chatbots with extensive customization options. These platforms offer pre-built models as well as the ability to train your models, depending on how specific the chatbot's tasks need to be.

Incorporating AI into web development is made easier with various tools and libraries designed to simplify the process. TensorFlow.js, for example, is a popular library that allows you to integrate machine learning models directly into your web browser, enabling you to run AI processes on the client side without the need for extensive server-side computations. This can be used for tasks like image or speech recognition directly within web applications, providing immediate interaction with users. Another useful tool is Amazon Web Services (AWS) AI services, which offer a range of AI capabilities that can be easily integrated into web applications, including language translation, text-to-speech, and visual recognition. These tools and libraries not only simplify the process of adding AI functionalities but also ensure that they are scalable and maintainable as part of larger web projects.

To illustrate the potential of AI in web development, consider the example of a retail website that uses AI to provide a personalized shopping experience. By analyzing user data such as past purchases, browsing history, and search patterns, the AI system offers personalized product recommendations, tailored promotional offers, and even dynamic pricing. Another example is a news aggregator website that uses natural language processing to analyze user preferences and reading habits, curating a personalized news feed for each user. Such implementations enhance user engagement and provide significant value by making the website more intuitive and responsive to individual needs.

By exploring these concepts and tools, you can begin to appreciate the profound impact AI can have on web development. Whether it's through enhancing user interaction, personalizing content, or automating routine tasks, AI opens up a spectrum of possibilities that can transform a standard website into a dynamic and intelligent platform. As you continue to explore the possibilities, remember that the goal of integrating AI into web development is to create more efficient, engaging, and personalized web experiences that meet the evolving expectations of users in this digital age.

4.4 AI AND PHOTOGRAPHY: ENHANCING IMAGES WITH AI

The advent of AI in photography has transformed the way we capture, edit, and perceive images. Utilizing AI technologies to enhance photographs is not simply about applying filters or adjustments; it's about intelligently analyzing and improving photos to achieve results that were once only possible with professional editing skills. AI in photography encompasses a range of techniques from basic enhancements like adjusting lighting and color balance to more complex edits such as object removal, style transfer, and even upscaling images to higher resolutions without losing quality.

At the core of AI-driven image enhancement is the ability of algorithms to understand and interpret the content of a photograph. This understanding enables AI to make context-aware enhancements that go beyond generic adjustments. For instance, AI can detect the sky in a landscape photo and enhance its blue hue without affecting the rest of the image. Similarly, it can recognize faces and automatically enhance features such as the eyes and skin, making the subject look their best. These intelligent adjustments are made possible through deep learning models that have been trained on thousands of images, allowing them to learn the nuances of what makes a photo visually appealing.

For those eager to explore AI-powered photo editing, numerous tools and applications make this technology accessible to everyone, regardless of their photography skills. Applications like Adobe Photoshop's AI-enhanced features and Luminar AI are designed to simplify complex editing tasks. These tools use AI to analyze your photos and suggest edits based on the content and type of photography. For example, if you upload a portrait, the AI might suggest enhancements to the skin tone or background blur, tailoring its suggestions to improve the specific type of photo.

To get started with enhancing your images using AI, follow this simple step-by-step tutorial using any AI photo editing software: First, upload your photo into the application. Most AI photo editors will automatically analyze the image and may even suggest initial edits right away. Explore the different AI features available —such as auto-enhance, which adjusts exposure, contrast, and color with a single click. Experiment with more targeted adjustments, like AI sky enhancement or AI portrait enhancer, depending on your photo's subject matter. Most tools also allow for manual fine-tuning, so you can adjust the intensity of the AI's enhancements to match your personal preference.

While the capabilities of AI in photography are impressive, it's important to approach AI-enhanced photography with an understanding of the ethical implications. As AI tools become more powerful, they make it easier to manipulate images in ways that can distort reality. For example, AI can be used to create deepfakes or alter images in misleading ways. Photographers and editors need to be mindful of the impact that significantly altered images can have on perceptions and beliefs. Ethical considerations should guide the use of AI in photography, ensuring that enhancements improve the aesthetic quality of images without deceiving viewers about the nature of the content. This ethical approach is essential in maintaining trust and authenticity in photography, particularly in fields like photojournalism where integrity is paramount.

The integration of AI into photography is just one example of how this technology is reshaping creative fields, making sophisticated editing techniques more accessible and opening up new possibilities for artistic expression. As AI continues to evolve, the line between technology and art becomes increasingly blurred, creating exciting opportunities for innovation in how we capture and present images. Whether you are a professional photographer looking to refine your workflow or a hobbyist eager to explore the potential of your photos, AI offers tools that can transform your approach to photography, making it easier to achieve visually stunning results.

4.5 BUILDING A SIMPLE AI TO PLAY YOUR FAVORITE GAMES

The allure of video games lies not only in their entertainment value but also in the fascinating world of artificial intelligence that often powers them. AI in gaming has revolutionized how games are played and developed, introducing levels of complexity and dynamism that mimic human-like interaction and intelligence. For beginners eager to explore the intersection of AI and gaming, developing a simple AI to play a video game can be an incredibly rewarding project.

When selecting a game for your AI project, it's essential to choose one that matches your current programming skills while still offering enough challenge to facilitate learning and growth. Start with something simple, such as classic games like Tic-Tac-Toe or Pong. These games have clear rules and limited outcomes, making them excellent choices for your first AI gaming project. The simplicity of these games allows you to focus more on the AI development aspects without getting overwhelmed by complex game mechanics. Once you understand the basics, you can gradually move to more complex games, like platformers or strategy games, where AI can make decisions based on a wider array of inputs and scenarios.

Developing an AI game agent, or a computer-controlled player, involves several steps. First, you need to define what the AI agent is supposed to do. In the context of a simple game like Tic-Tac-Toe, the AI's role is to analyze the board and decide the best move based on the current state of the game. To achieve this, you can use a decision-making algorithm like the Minimax algorithm. This algorithm plays out all possible moves in the game to determine which move will lead to the best outcome. It considers the minimization of the possible benefit for the opposing player, aiming to maximize the chances of the AI winning. Implementing this involves setting up the game logic to allow the AI to evaluate the game board, make predictions about potential outcomes, and decide on the optimum move.

The development of your AI game agent should also include a testing phase where you can play against the AI to see how it performs. This testing phase is crucial as it provides firsthand insight into how your AI reacts to different game situations. You'll likely need to go back and tweak the AI's decision-making process, especially if you find it's making predictable or poor decisions. The beauty of this phase lies in the iterative process of refining your AI, which involves continuously adjusting and improving the AI logic based on game outcomes and performance.

For those looking to dive deeper into AI gaming, participating in AI gaming competitions can provide extensive learning opportunities. Numerous online platforms host AI challenges, where developers pit their AI systems against one another in specific games. These competitions are not only a great way to test your AI in a competitive environment, but also a fantastic way to learn from others in the community. Observing how different AI agents perform and analyzing the strategies used by top competitors can provide insights that you can apply to your own AI projects. Furthermore, many of these competitions provide forums where participants discuss tactics and share advice, making them an excellent resource for beginners seeking guidance and feedback.

Engaging in such projects and competitions exposes you to the practical applications of AI in gaming, cementing your understanding of both the technological aspects and the strategic thinking required to create AI that can effectively play and compete in video games. As you progress, the experience gained can be instrumental in tackling more complex AI projects, whether in gaming or other fields that require intelligent decision-making systems. This hands-on approach enhances your technical skills and sparks creativity, as each game presents unique challenges and opportunities for AI application.

4.6 AI IN CONTENT CREATION: WRITING ARTICLES WITH AI

The landscape of content creation is undergoing a transformative shift with the integration of artificial intelligence. AI's capability to generate written content is not just about automating tasks; it's about enhancing the creativity and efficiency of content creators. From drafting marketing copy to composing full-length articles, AI tools are increasingly becoming indispensable assets in the writer's toolkit. These AI systems are designed to understand and manipulate language in ways that mimic human writing, offering a blend of speed and scale that can significantly amplify a content creator's output.

AI's role in content creation primarily revolves around generating, augmenting, and refining text. This technology is based on sophisticated algorithms capable of analyzing vast datasets of text to learn linguistic patterns, styles, and structures. By training on diverse sources of text, AI models like OpenAI's GPT (Generative Pre-trained Transformer) have achieved remarkable proficiency in generating coherent and contextually relevant text that closely resembles human-written content. The applications are vast — from helping journalists draft articles based on data inputs to assisting novelists in overcoming writer's block by suggesting narrative ideas or dialogue.

When it comes to tools that facilitate AI-assisted writing, several options stand out due to their accessibility and advanced capabilities. Tools like Grammarly use AI not only to correct grammar and spelling but also to enhance clarity and tone, offering suggestions that improve the overall quality of the writing. Another powerful tool, Jasper, offers features that enable users to produce content on a wide range of topics efficiently, providing flexibility in style and format based on the user's needs. These tools are designed to be intuitive, allowing even those with minimal technical skills to leverage AI in their writing processes effectively.

For those looking to harness AI for writing articles, the process involves more than just understanding how to operate these tools; it requires a strategy to integrate AI effectively into your content creation workflow. Start by defining the scope and purpose of your content to guide the AI in generating relevant text. For instance, if you are writing an article on health and fitness, you might use an AI tool to generate content on recent trends or data-driven insights, which you can then refine and expand based on your expertise and research. It's also beneficial to iteratively review and edit the content generated by AI. While AI can provide a solid base or draft, the human touch is essential to ensure the content resonates with the audience and maintains a genuine voice. Regularly updating the AI tool's knowledge base and parameters will keep the outputs fresh and aligned with current trends and linguistic preferences.

Navigating the ethical landscape of using AI in content creation is crucial. As AI tools become more sophisticated, the lines between human-generated and machine-generated content blur, raising questions about authenticity and intellectual property. Ensuring transparency about the use of AI in content creation is fundamental; readers should be aware if the content they are consuming is generated by AI. Additionally, there is a growing need to address the potential for misinformation, as AI could be used to create convincing yet factually incorrect or misleading content. Establishing ethical guidelines and rigorous fact-checking protocols is essential to mitigate these risks and ensure that AI is used responsibly in content creation.

As AI continues to evolve and become more integrated into the field of content creation, embracing these technologies offers a pathway to enhanced creativity and productivity. However, it also necessitates a commitment to ethical practices and continuous learning to adapt to new developments and possibilities in AI. This exploration into AI-driven content creation not only expands our toolkit as writers and creators but also challenges us to rethink the dynamics of creativity and the role of technology in artistic expression. By understanding and leveraging AI responsibly, we can unlock new dimensions of creativity that harmonize human intuition with machine efficiency, paving the way for innovations that enrich our narratives and engage our audiences more deeply.

In summary, this chapter has unveiled the exciting potential of AI in revolutionizing content creation, from drafting articles to refining written content with unprecedented efficiency. As we transition into the next chapter, we will explore further how AI is reshaping other creative fields, continuing our exploration of this transformative technology in modern applications. The journey through AI's capabilities in content creation is just one facet of its broader impact on industries and professions, highlighting the importance of integrating technology with traditional practices to foster innovation and growth in an increasingly digital world.

CHAPTER FIVE

AI TOOLS AND TECHNOLOGIES

In the vast universe of artificial intelligence, the tools and platforms available are as varied and dynamic as the stars in the night sky. Each tool offers unique capabilities and opportunities for innovation, whether you're a budding enthusiast eager to dip your toes into AI waters or someone looking to implement a specific project. Among these shining stars, Google's suite of AI tools stands out, offering both the breadth and depth needed to explore and harness the potential of AI effectively. In this section, we will navigate the expansive realm of Google's AI resources, providing you with the knowledge to start leveraging these powerful tools in your projects.

5.1 EXPLORING AI WITH GOOGLE'S AI TOOLS

Google's AI offerings: An overview of the AI tools and resources provided by Google

Google has been at the forefront of AI development, creating tools that empower both novices and experts to bring their AI ideas to life. From machine learning frameworks like TensorFlow to cloud-based AI services via Google Cloud, the array of tools is designed to meet a wide spectrum of needs. TensorFlow, for example, is an open-source library that allows you to develop your machine-learning models with relative ease. It supports various applications, from beginner-friendly experiments to large-scale deployments. Additionally, Google AI provides various APIs such as the Vision API and Natural Language API, which facilitate the integration of advanced AI capabilities into applications without requiring deep knowledge of machine learning algorithms. These APIs can analyze images, comprehend and generate text, translate languages, and more, making them incredibly versatile tools for developers looking to incorporate AI functionalities.

Getting started with Google AI: Step-by-step instructions on how to access and use Google's AI tools for beginners

For beginners eager to start their journey with Google's AI tools, the first step is often the most daunting. However, Google has made this initiation as seamless as possible. Most of Google's AI tools, like TensorFlow and various AI APIs, are well-documented with comprehensive guides and tutorials. To get started, you would typically begin by setting up an account on Google Cloud, which provides access to all the AI tools offered by Google. This platform hosts the tools and offers extensive resources such as tutorials and community forums that can help you navigate your initial projects. For instance, to use TensorFlow, you can download and install it on your computer, followed by running a simple program to ensure it's set up correctly. Google Colab, another valuable resource, offers a browser-based coding environment that comes pre-installed with TensorFlow and other libraries, allowing you to write and execute code without any local setup.

Practical applications for Google's AI tools: Examples of how beginners can utilize Google AI in their projects

The practical applications of Google's AI tools are limited only by one's imagination. For beginners, a simple project might involve using the Vision API to build an image recognition system that can categorize photos. For instance, you could create a program that identifies different plant species from images. This project would involve feeding the API with images and using the model it provides to classify these images into predefined categories. Another interesting project could be employing the Natural Language API to develop a sentiment analysis tool, which assesses the emotion conveyed in pieces of text. Such a tool could analyze customer reviews to determine overall sentiment about a product, providing valuable feedback to businesses. These projects serve as excellent learning experiences and have real-world applications that can add significant value in various sectors.

Tips for maximizing the benefits of Google AI: Advice on how to effectively leverage Google's AI resources for personal learning and development

To truly benefit from Google's AI tools, it is crucial to approach them with a strategy that fosters both learning and practical application. One effective approach is to engage actively with the community and resources available. Participating in forums and following tutorials specific to Google AI tools can provide insights and help solve any challenges you encounter. Additionally, Google often offers free trials and credits for beginners to experiment with their cloud-based services, which you can take advantage of to explore different tools without incurring initial costs. It's also beneficial to keep your projects aligned with your learning objectives; start with simple, manageable projects to build your confidence and understanding before progressing to more complex applications. Finally, always keep an eye on new releases and updates from Google, as the landscape of AI tools is continually evolving, and staying updated can provide you with more advanced tools and functionalities to enhance your projects and skills further.

Navigating the realm of Google's AI offerings opens up a world of possibilities. With the right tools, guidance, and a dash of creativity, you can begin to tap into the transformative potential of AI, making what once seemed like daunting technologies accessible and manageable. Whether you're looking to implement a specific project or simply explore what's possible with AI, Google's tools provide a robust foundation for your endeavors, ensuring you have the support and resources needed to succeed in this exciting field.

5.2 LEVERAGING OPENAI'S PLATFORMS FOR BEGINNERS

OpenAI stands as a beacon in the AI community, founded with the mission to ensure that artificial intelligence benefits all of humanity. The organization is committed to democratizing AI technology, making powerful tools accessible to everyone, from seasoned researchers to curious beginners. Understanding and using OpenAI's platforms can be a transformative experience for those new to AI, offering a unique opportunity to engage with cutting-edge technologies.

OpenAI provides several user-friendly platforms that are particularly suited for beginners. One of the most prominent is the OpenAI Gym, which offers a suite of environments that simulate various tasks or games, making it an ideal testing ground for writing and assessing algorithms. These simulations range from simple physical tasks to complex games, each designed to challenge different aspects of AI algorithms. Another significant offering is the OpenAI API, which provides access to models trained on a diverse range of internet text. With this API, beginners can implement advanced natural language processing tasks, from sentiment analysis to content generation, without needing to train their own models from scratch. The accessibility of these platforms is pivotal for those who are just starting out, as it allows them to experiment with and learn about AI without the overhead of developing complex models themselves.

Engaging with OpenAI's technologies can be both educational and inspiring. For beginners, a practical project to start with could be using the OpenAI Gym to create an agent that learns to play a simple game, such as CartPole, where the goal is to balance a pole on a moving cart. This project introduces beginners to the concepts of reinforcement learning, a type of machine learning where an agent learns to make decisions by trial and error, receiving rewards or penalties. The process involves defining the strategy (policy) that the agent uses to decide its actions, which is refined as the agent learns from continuous interaction with the environment. Another project could involve using the OpenAI API to develop a chatbot that can converse on a wide range of topics. This not only helps in understanding natural language processing but also provides practical experience in integrating and utilizing AI in real-world applications.

However, while exploring these possibilities, it is crucial to understand OpenAI's policies and the limitations of its platforms. OpenAI promotes safe and responsible AI usage, and its API comes with usage policies that restrict the generation of harmful content and ensure ethical usage. These policies are in place to prevent misuse of the technology and to foster a community that uses AI ethically. Beginners must familiarize themselves with these guidelines to ensure their projects align with OpenAI's principles. Moreover, while OpenAI's tools are powerful, they are not omnipotent. They have limitations in terms of the complexity of tasks they can handle and the type of data they have been trained on. For instance, while the OpenAI API is excellent for text-based tasks, it may not be suitable for tasks requiring visual understanding or other types of data.

Navigating the capabilities and resources offered by OpenAI can open up a world of possibilities for those new to AI. By starting with simple projects and gradually increasing complexity, beginners can gain confidence and develop an in-depth understanding of how AI works and its potential applications. The journey with OpenAI's platforms can be as broad and deep as one's curiosity permits, with each project providing new insights and challenges. Whether it's mastering the basics of machine learning through interactive simulations or exploring the frontiers of natural language processing, OpenAI provides the tools and resources to foster learning and innovation. As you continue to explore these technologies, remember that each step taken is a building block in understanding the broader implications and opportunities of AI.

5.3 AI DEVELOPMENT KITS FOR NON-CODERS

Artificial Intelligence, once an intimidating field reserved for those with deep coding expertise, has now expanded its reach, thanks to AI development kits specifically designed for non-coders. These kits are a boon for beginners who are eager to explore AI but may not have the technical background typically associated with the field. By providing a user-friendly interface and pre-built modules, these kits lower the barriers to entry, allowing anyone with curiosity and a creative mindset to embark on AI projects.

AI development kits for beginners are crafted to simplify the complexities of AI technologies. These kits often come with drag-and-drop functionalities, pre-coded modules, and comprehensive guides that walk you through each step of the process. For example, kits like Google's AIY Projects offer everything you need to start building AI projects, such as voice recognition and image processing devices, right out of the box. These kits not only include the physical components but also access to software tools that let you program and control these devices without writing extensive lines of code. This approach demystifies AI and makes the learning process much more engaging and manageable.

When selecting the right AI kit, it's crucial to consider a few key factors to ensure that the kit meets your needs and interests. First, identify your primary area of interest within AI—be it robotics, natural language processing, or something else. Different kits focus on different aspects of AI, so aligning the kit with your interests will make the learning process more enjoyable and effective. Next, consider the level of support and community around the kit. A robust community and good support resources can greatly enhance your learning experience, providing help when you encounter challenges and inspiration through community projects. Lastly, evaluate the scalability of the kit. While your initial projects might be simple, choosing a kit that can handle more complex applications as your skills develop can offer better long-term value.

For your first project with an AI development kit, a smart approach is to start simple. One potential project could involve building a voice-operated device using a kit like the Google AIY Voice Kit. This project could teach you the basics of voice recognition technology and its application in creating interactive devices. Here's a step-by-step guide to get you started:

1. Assemble the Kit: Begin by assembling the physical components of the kit. This usually involves connecting various modules like microphones, speakers, and processors. Following the detailed instructions provided with the kit will help you correctly set up the hardware.
2. Install the Software: Download and install any necessary software from the kit's resource page. This software will enable you to program the device and customize its functionalities.
3. Customize Voice Commands: Use the graphical programming interface to set up basic voice commands. For instance, you could program the device to turn on a light or play music in response to specific commands.
4. Test and Refine: Test the functionality of your device by experimenting with different commands and scenarios. Refine the commands based on performance, and

troubleshoot any issues with the help of online forums or support resources.

5. Explore Further: Once you're comfortable with the basics, explore more complex applications, such as integrating your device with IoT systems or adding additional sensors to enhance its capabilities.

To further your learning and development with AI kits, numerous resources are available. Online platforms like Coursera and Udemy offer courses specifically designed for beginners using these kits. These courses often include hands-on projects and tutorials that can deepen your understanding of both the hardware and software aspects of AI. Additionally, joining online forums and communities, such as those on Reddit or specialized websites, can provide ongoing support and inspiration as you continue to explore and learn. Engaging with these communities allows you to share your projects, receive feedback, and connect with others who are also on their AI learning path.

By starting with a user-friendly AI development kit, you can gradually build your understanding and skills in this fascinating field. The hands-on experience gained from working directly with AI technologies demystifies the complexities of the field and opens up a world of creative possibilities. Whether your interest lies in creating practical devices or exploring theoretical concepts, AI development kits provide a solid foundation for both learning and innovation.

5.4 SIMPLE AI MODELS YOU CAN TRAIN AT HOME

Understanding AI Models: A Brief Introduction to What AI Models Are and How They Learn

Artificial intelligence models, in essence, are systems that have been trained to process input data and make decisions or predictions based on that data. Imagine teaching your pet to fetch; you reward them when they succeed, and over time, they learn what to do. Similarly, AI models learn from data. They adjust their internal parameters (think of these as their 'thought processes') based on the feedback they receive from their performance on training data. This method, often involving a lot of trial and error, refines their ability to perform tasks such as recognizing images, understanding spoken words, or predicting stock market trends.

The learning process of an AI model is iterative. Initially, the model makes predictions using its initial settings, which are usually random. These predictions are then compared against the correct answers, and the difference between the prediction and the correct answer (the error) is used to adjust the model's parameters. This process is repeated numerous times over thousands, sometimes millions, of examples. Over time, the model's predictions become increasingly accurate as it 'learns' from the accumulated data. This is akin to practicing a musical instrument - the more you practice, the better you get.

AI learning is primarily categorized into supervised learning, where the model learns from examples that have known answers, and unsupervised learning, where the model tries to find patterns in data without prior knowledge of answers. For beginners, supervised learning tends to be more straightforward and is a practical starting point for building AI models at home.

Tools for Training AI Models: Listing Accessible Tools and Platforms That Allow Beginners to Train AI Models

For those new to AI, the thought of training an AI model might seem daunting. Fortunately, several tools and platforms make this process accessible and straightforward. Platforms like Scikit-learn, Keras, and TensorFlow provide user-friendly interfaces and comprehensive libraries that simplify the coding required to build and train AI models. These platforms are supported by extensive documentation and active communities, where beginners can find tutorials, example projects, and advice.

Scikit-learn, for example, is particularly well-suited for beginners due to its simplicity and the wide range of algorithms it supports. It provides simple functions that can be used to pre-process data, split it into training and testing sets, train a model, and assess its performance. Keras, working on top of TensorFlow, offers a higher-level, more intuitive set of tools that allow for easy construction of neural networks. It abstracts many of the complex details of building neural networks, making it easier for non-experts to build powerful AI models.

Training Your First Model: A Beginner-Friendly Guide to Training a Simple AI Model on Your Own Data

Training your first AI model can be an exciting experience, especially when you see it start to make accurate predictions. A good starting project could be a simple image classification model where the task is to identify whether a photo contains a cat or a dog. For this, you can use TensorFlow combined with Keras, which provides a straightforward way to set up the layers of your neural network.

First, you'll need a dataset, which in this case could be a collection of labeled images of cats and dogs. You can find such datasets easily available online, or you could even create your own by taking photos of pets and labeling them. Once you have your dataset, the next step is to pre-process the data to ensure it is in a format that the model can work with effectively. This typically involves resizing the images to a consistent shape and normalizing the pixel values.

Next, you would define your model. In Keras, this can be done by stacking layers in a sequential model, starting with convolutional layers that help the model "see" the different features in the images, followed by pooling layers that reduce the dimensionality of the data, and finally, dense layers that predict the output. After defining the model, you compile it, specifying the loss function and the optimizer, which are essentially the rules the model follows to minimize error.

The final step is to train the model by calling the 'fit' method, passing in your training data. As the model trains, you can watch its performance improve as it processes the data over several iterations, learning from each image.

Applying Your Trained Model: Ideas for How to Apply Your Newly Trained AI Model in Practical Scenarios

Once your model is trained, the possibilities for application are vast. For instance, your image recognition model can be integrated into a mobile app that helps users organize their photo libraries by automatically categorizing images of pets. Alternatively, it could be used in a wildlife camera that sends notifications when a cat or dog is detected, potentially keeping an eye on pet safety or monitoring for stray animals.

Another exciting application could be integrating your model with a home assistant device to create an interactive tool for pet owners. For example, the device could use the model to identify pets from a camera feed and then provide information or perform specific tasks based on the pet it sees, like playing music if it recognizes the cat in a room, assuming it helps keep the pet calm.

These projects not only provide practical uses for your AI model but also offer the satisfaction of seeing your creation solve real-world problems. Each application serves as a stepping stone to more complex and impactful AI projects, expanding your skills and opening up new possibilities in the world of artificial intelligence.

5.5 USING CLOUD SERVICES FOR AI PROJECTS

Cloud-based AI services provide a compelling advantage, especially for beginners venturing into the realm of artificial intelligence. These services offer on-demand access to powerful computing resources and advanced AI tools, which can be a significant boon for those who might not have the extensive hardware typically required for AI projects. Essentially, cloud services democratize the ability to engage with AI by removing the barrier of high initial investment in physical infrastructure. This means you can start experimenting with and deploying AI models without the need to purchase expensive GPUs or servers. This accessibility is particularly beneficial for beginners who are looking to experiment and learn without committing a substantial amount of resources upfront.

One of the most significant benefits of using cloud-based AI services is scalability. These services allow you to start small, testing ideas and models with minimal resources, and scale up as your needs grow, without the need for substantial upfront investments. This flexibility is crucial for learning and innovation in AI, where you might need to iterate several times to refine your models and approaches. Moreover, cloud services often come with built-in tools and pre-trained models, which you can use to accelerate your development process. These tools can perform a variety of tasks, from data preprocessing and model training to deployment and monitoring, all within the same ecosystem. This integration can significantly simplify your AI projects, allowing you to focus more on learning and less on managing infrastructure.

Navigating the landscape of available cloud AI services can be overwhelming given the plethora of options. The leading players in this field include AWS (Amazon Web Services), Microsoft Azure, and Google Cloud, each offering a suite of AI services that cater to different needs. AWS, for example, provides a comprehensive set of machine learning services and tools that are designed to be accessible to both beginners and experienced developers. Azure offers services like Azure Machine Learning, which provides tools to accelerate the end-to-end machine learning lifecycle. Google Cloud's AI platform integrates seamlessly with other Google services and provides robust tools for job training and prediction. Each platform has its strengths and specialties, and the choice between them often depends on specific project requirements, such as the type of AI applications you intend to develop, or integration with other tools and services you might already be using.

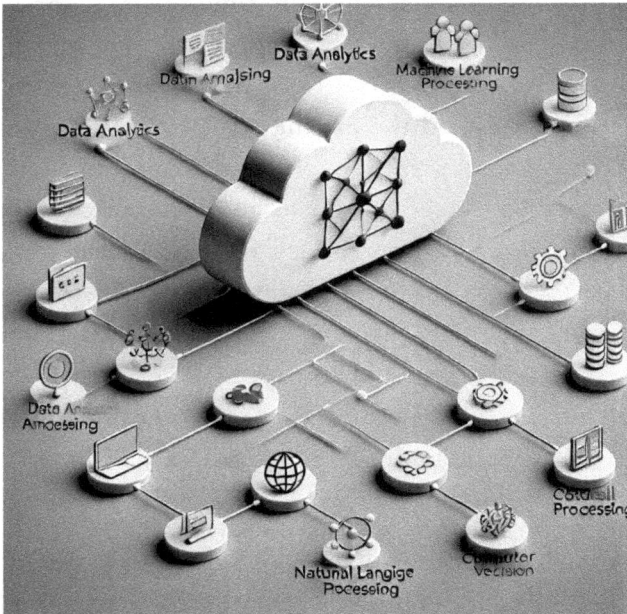

To start using cloud AI services, the initial steps typically involve setting up an account on your chosen cloud platform. This process is usually straightforward, involving registration and possibly setting up billing arrangements, as many cloud services operate on a pay-as-you-go basis. After setting up your account, the next step is to familiarize yourself with the environment and the specific AI tools it offers. Most cloud services provide extensive documentation and tutorials, which are invaluable resources as you begin. For instance, you might start by exploring a tutorial on how to deploy a simple machine-learning model using the platform's tools. This hands-on approach not only helps solidify your understanding of the cloud services but also gives you practical experience in deploying and managing AI models.

Managing costs is a crucial aspect of using cloud AI services, especially for beginners who are just exploring and may not have large budgets. Cloud platforms typically charge based on the resources you use, such as compute time or data storage, which means costs can vary widely depending on the scale and complexity of your projects. To manage costs effectively, it's important to understand the pricing structure of the services you are using. Most cloud providers offer calculators to estimate costs based on expected usage, which can help you budget accordingly. Additionally, many platforms offer free tiers or trial periods with access to certain services at no cost, which can be a great way to learn without incurring expenses. However, it's important to monitor your usage to avoid unexpected charges, especially as you scale your projects. Setting up alerts to notify you when you are approaching your budget limit can be a helpful tool to keep your costs under control.

In summary, cloud services open up a world of possibilities for AI projects, providing the tools and resources needed to build, train, and deploy AI models effectively. Whether you're a beginner looking to learn and experiment or someone aiming to develop complex AI solutions, cloud AI services offer the scalability, flexibility, and support needed to succeed in your endeavors. As you continue to explore these services, remember that each step forward enhances your understanding of both the potential and challenges of AI, guiding you through a landscape rich with opportunities for innovation and growth.

5.6 UNDERSTANDING AND USING AI APIS FOR PROJECTS

Navigating the realm of Artificial Intelligence (AI) can seem like a complex endeavor, particularly when you start exploring the practical aspects of implementing AI in real-world applications. One of the most accessible ways to integrate AI into your projects is through the use of AI Application Programming Interfaces (APIs). AI APIs act as bridges, allowing you to tap into powerful AI capabilities without the need to build models from scratch. Essentially, these APIs provide ready-made intelligence that you can integrate into your applications, whether you're developing a new app or enhancing existing software.

AI APIs, or Application Programming Interfaces for Artificial Intelligence, are tools that allow your applications to interact with an extensive range of AI functionalities hosted on remote servers. These APIs are made available by major tech companies and AI research organizations, and they cover a broad spectrum of AI capabilities, from natural language processing and speech recognition to image analysis and decision-making algorithms. By using these APIs, you can add complex AI features to your projects without the deep expertise typically required to develop AI from the ground up. For example, if you're building an app that requires the ability to recognize and interpret human speech, instead of creating your own speech recognition system, you can integrate an AI API like Google's Cloud Speech-to-Text. This saves time and resources and ensures you're leveraging some of the most advanced AI technologies available.

Choosing the right AI API for your project is crucial and depends largely on the specific needs of your application. Start by defining the functionality you need. Are you looking to translate text? Analyze sentiments in customer reviews? Recognize objects in images? Once you've pinpointed the exact AI capabilities you need, you can explore different AI APIs that offer those functionalities. Each API comes with its own set of features, limitations, and cost structures, so it's important to compare these aspects to find the one that best fits your project's requirements. Additionally, consider the ease of integration and the level of community and developer support available, as these can significantly affect how smoothly you can implement and maintain the API in your application.

Integrating an AI API into your project typically involves several key steps. First, you'll need to sign up for the API service and obtain an API key, which is used to authenticate your requests to the API. Most AI APIs have comprehensive documentation that includes instructions on how to make requests to the API and handle responses. For instance, if you are using an API for language translation, the documentation will guide you on how to format your text input and will detail the structure of the response you'll receive, which will include the translated text. Testing is an integral part of the integration process. Initially, make calls to the API using test data to ensure that it's returning the expected results. Once you're satisfied with the test outcomes, you can proceed to integrate the API calls into your application code, replacing the test data with dynamic input from your app users.

To illustrate the power and versatility of AI APIs, let's consider some project examples where these tools can be effectively utilized. One simple yet impactful project could involve integrating a natural language processing API to develop a customer service chatbot. This chatbot could analyze incoming customer queries, understand the sentiment behind them, and provide appropriate responses, enhancing customer interaction without human intervention. Another project could be an educational app that uses an image recognition API to help students learn about wildlife. By uploading photos of animals, the app could provide information about the species, its habitat, and conservation status, creating an interactive learning experience.

As we wrap up this exploration of AI APIs, it's clear that these tools offer a gateway to incorporating sophisticated AI functionalities into your projects with relative ease. By understanding how to select, integrate, and utilize these APIs, you can significantly enhance the capabilities of your applications, making them more intelligent, interactive, and user-friendly. Whether you're a hobbyist looking to experiment with AI or a developer aiming to build advanced AI-driven applications, AI APIs provide the resources you need to bring your projects to life.

As this chapter concludes, we've seen how AI APIs can transform your projects by adding cutting-edge intelligence in a manageable way. This exploration demystifies AI and opens up a spectrum of possibilities for innovation and creativity in your applications. As we move forward, the next chapter will delve into another fascinating aspect of AI, ensuring that your journey through this book continues to be engaging and enlightening.

CHAPTER SIX

AI FOR CAREER ADVANCEMENT

As you venture further into the realms of artificial intelligence, the horizon of career opportunities expands before you, offering myriad pathways illuminated by the glow of technological progress. The digital age thrives on innovation and adaptability, qualities that AI not only embodies but also demands from its enthusiasts. If you're considering a career in this vibrant field, understanding what skills are in high demand, how to acquire them, and how to effectively showcase them to potential employers is crucial. This chapter is designed to guide you through these steps, ensuring you are well-prepared to meet the evolving demands of the job market in AI.

6.1 AI SKILLS THAT EMPLOYERS ARE LOOKING FOR

In the fast-evolving landscape of AI, certain skills stand out for their high demand among employers. These include proficiency in machine learning, knowledge of neural networks, expertise in natural language processing, and the ability to work with big data platforms. Additionally, programming skills in languages such as

Python, R, and Java are often essential. Understanding these technologies and their applications in real-world scenarios can significantly enhance your employability in various industries, from tech startups to multinational corporations seeking to integrate AI into their operations.

Building a path to acquiring these skills involves a combination of self-study and formal education. Online platforms like Coursera and Udacity offer courses designed by industry experts that cover a range of topics from basic programming to advanced AI concepts. These platforms often provide practical, project-based learning experiences that are invaluable for deepening your understanding and skill set. For those who prefer a more structured approach, pursuing degrees or certifications in data science, computer science, or AI from reputed educational institutions can provide comprehensive training along with the credentials that many employers find attractive.

Showcasing your AI skills to potential employers goes beyond listing them on your resume. Developing a portfolio of projects that demonstrate your expertise and the impact of your work can be far more persuasive. Consider creating a personal website or a digital portfolio on platforms like GitHub to display your projects. This could include anything from a machine-learning model that predicts consumer behavior to a natural language processing project that analyzes and interprets customer feedback. Each project should be accompanied by a clear description of your role, the technologies used, and the outcome, highlighting how your contribution drove success.

Continuous skill development is crucial in a field as dynamic as AI. The industry evolves rapidly, with new tools, techniques, and best practices emerging regularly. Staying updated requires an ongoing commitment to learning. Participating in workshops, webinars, and conferences, as well as staying engaged with the latest research and publications in AI, can help you keep pace with new developments. This enriches your skill set and demonstrates to employers your dedication to staying at the forefront of technological advancements.

Interactive Element: AI Skills Self-Assessment Quiz

Take this interactive quiz to evaluate which AI skills you already possess and identify areas where you might need further development. This self-assessment can help you tailor your learning path and ensure you are focusing on the skills that will most effectively boost your career in AI.

Navigating a career in AI demands a proactive approach to learning and an openness to continually adapting your skills. As you build and showcase your capabilities, you are not only preparing yourself for current opportunities but are also paving the way for future advancements in your career. The journey through AI is as much about understanding and leveraging technology as it is about personal and professional growth. As such, your journey is shaped by both your command of technical skills and your strategic vision for their application in solving real-world problems.

6.2 HOW TO SHOWCASE YOUR AI PROJECTS ON YOUR RESUME

When preparing to apply for positions in the AI field, your resume not only serves as your first point of contact with potential employers but also as a showcase of your professional journey and capabilities. Especially in a field as dynamic and impact-driven as artificial intelligence, the projects you choose to highlight on your resume can significantly influence your career prospects. Selecting which AI projects to feature should be a strategic decision. Focus on projects that align with the job role you are targeting or those that demonstrate a breadth of skills and a depth of understanding. Projects that solved real-world problems or that were innovative in their approach are particularly compelling. They not only demonstrate your technical skills but also your ability to apply these skills effectively.

Describing your AI projects on your resume requires more than just listing your tasks. It involves crafting a narrative that highlights your role, the challenges you addressed, and the outcomes of your projects. Start by succinctly describing the problem or objective of the project. This sets the stage for understanding the context and significance of your work. Follow this by detailing the AI technologies and methodologies you employed, emphasizing any unique or advanced techniques that showcase your expertise. Be specific about your contribution to the project to articulate your role and responsibilities. Finally, describe the outcome of the project. Wherever possible, quantify these results, as tangible achievements are compelling evidence of your capabilities. For example, if your AI model improved the efficiency of a process, specify the percentage increase in efficiency or the time and cost savings achieved.

Incorporating visual and interactive elements into your resume can make your application stand out in the competitive AI job market. Consider including links to an online portfolio or project demos. This can be particularly effective for roles that involve machine learning, data visualization, or user interface design, where visual evidence of your skills can be more persuasive than text descriptions alone. Ensure that any linked content is professional, up-to-date, and accessible to those reviewing your application. This not only demonstrates your technical skills but also your professionalism and attention to detail.

Quantifying the impact of your projects is crucial in demonstrating the value you can bring to a potential employer. It translates your technical achievements into business or operational terms that hiring managers can appreciate. Whether it's through improving accuracy, and efficiency, or enabling new capabilities, showing how your work has led to measurable improvements can set you apart from other candidates. Use clear metrics and provide context to these figures to ensure they resonate with those reading your resume. For instance, instead of stating "reduced errors," specify by saying "reduced classification errors by 25% over six months, leading to a 40% decrease in customer complaints."

By thoughtfully selecting, describing, and quantifying your AI projects on your resume, you transform it from a mere document listing your experiences into a powerful tool that showcases your professional journey and achievements. This approach captures the attention of potential employers and provides a clear narrative of your growth and capabilities in AI, setting the stage for discussions in interviews and beyond. As AI continues to evolve, so too should the way we present our professional selves, ensuring we capture the full spectrum of our skills and the impact of our work.

6.3 NETWORKING IN THE AI COMMUNITY: TIPS AND TRICKS

In the rapidly evolving field of artificial intelligence, networking is not just about building contacts; it's about creating a community that fosters learning, innovation, and opportunities. For anyone aspiring to make significant strides in AI, understanding the immense value of networking is crucial. It's a dynamic way to stay informed about the latest developments, gain insights from experienced professionals, and even discover potential career opportunities. Networking in AI can help you bridge the gap between academic knowledge and real-world applications, providing you with a broader understanding of how AI technologies are being utilized across different sectors.

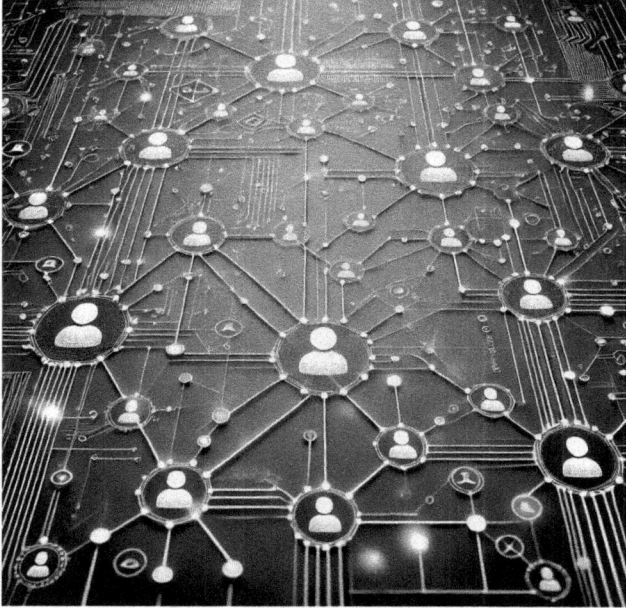

Finding vibrant AI communities where beginners can start networking might seem daunting at first, but numerous platforms cater to varying levels of expertise and interests. Online forums and communities such as Stack Overflow, GitHub, and Reddit offer spaces where you can ask questions, share projects, and learn from others' experiences. These platforms are treasure troves of information and provide a sense of how diverse the field of AI really is. For those who prefer face-to-face interaction, look for local meetups, workshops, or conferences focused on AI and related fields. Organizations like Meetup.com often host events that bring together AI enthusiasts from various backgrounds.

Attending these events can provide you with direct access to industry insiders and innovators who can provide guidance and potentially influence your career path positively.

When it comes to best practices for making meaningful connections within the AI community, it's essential to be both proactive and considerate. Start by clearly defining your goals for networking. Are you seeking mentorship, looking for collaborators, or hoping to find job opportunities? With your objectives in mind, you can more effectively engage with individuals and discussions that align with your goals. When interacting with others, whether online or in person, remember that the key to effective networking is reciprocity. Be ready to offer something of value, be it your unique perspective, expertise, or even support for others' projects. This approach not only enriches the community but also establishes you as a thoughtful and engaged member, which can lead to more meaningful and beneficial connections.

Leveraging social media platforms like LinkedIn and Twitter is another strategic way to enhance your networking efforts. These platforms allow you to follow and interact with AI professionals and thought leaders, join AI-focused groups, and share your own insights and projects. When using these platforms, it's important to maintain a professional profile that reflects your interest and expertise in AI. Regularly update your profile with any new projects, courses, or skills you acquire. Engage with content posted by your connections by commenting with thoughtful insights or sharing relevant articles and research papers. This keeps you visible in your network and helps build your reputation as someone knowledgeable and passionate about AI.

Through effective networking, you can gain a more in-depth understanding of the field, learn from the experiences of others, and open doors to opportunities that might otherwise remain unreachable. Remember, every conversation could lead to a new perspective or opportunity, so approach each interaction with openness and curiosity. By actively participating in AI communities and leveraging platforms to connect with professionals, you can significantly enrich your learning journey and career prospects in artificial intelligence.

6.4 TRANSITIONING YOUR CAREER TO AI AND TECH

If you're contemplating a shift towards a career in artificial intelligence and tech, the first step is to thoroughly evaluate your existing skill set to understand how it aligns with the demands of this dynamic field. Many skills you possess from previous roles, whether in project management, data analysis, software development, or even critical thinking, can serve as a solid foundation for a career in AI. For instance, if your background includes quantitative analysis, you already have a foothold in understanding data-driven decision-making, a core component of AI applications. Similarly, experience in software development directly translates to skills needed for coding AI algorithms. Begin by listing your current skills and then aligning them with the requirements of roles in AI that interest you. This exercise not only helps you see the valuable skills you already possess but also frames your mindset positively as you consider this transition.

Identifying gaps in your skills is equally crucial, as it helps you pinpoint the areas where you need further development to effectively contribute to AI projects. Begin by researching job descriptions for AI roles that intrigue you and note the specific skills and qualifications these positions require. Commonly, you might find gaps in areas like programming languages specific to AI such as Python or R, machine learning techniques, or familiarity with AI platforms and tools. Once these gaps are identified, the next step is to seek resources to fill them. Online courses are a fantastic way to do this. Platforms like edX, Coursera, and Udacity offer courses tailored to all levels of expertise in AI and machine learning. These platforms often feature courses developed by universities or tech companies, providing not only education but also credibility to your learning efforts. Additionally, many of these courses include hands-on projects that help you apply what you've learned in practical scenarios, enhancing your understanding and appeal to potential employers.

Creating a detailed plan for transitioning into an AI career involves setting clear, achievable goals and timelines. Start by defining where you want to be in your AI career within a specific timeframe, say one or two years. Break this long-term goal into smaller, manageable objectives such as completing specific courses, attending workshops, or building projects. Setting regular milestones for these objectives will help you maintain a steady pace and keep track of your progress. It's also beneficial to engage with a mentor who has expertise in AI. A mentor can offer guidance tailored to your personal career goals, provide insights into the industry, and even help you network with professionals in the field. Your plan should be flexible enough to accommodate new learning opportunities and changes in the AI landscape, which can influence the skills you need to focus on or the roles that become available.

Inspirational stories of individuals who have successfully transitioned into AI careers can serve as powerful motivation during your journey. Consider the story of a former financial analyst who leveraged her expertise in data analysis to move into a machine-learning role at a tech company. She began her transition by taking online courses in machine learning and participating in Kaggle competitions to apply her new skills. Her persistent effort expanded her skill set and built a portfolio of projects that showcased her capability to apply machine-learning techniques in real-world scenarios. Another example is a software developer who transitioned to an AI role within the same company. He took the initiative to learn AI by enrolling in an AI certification program and started contributing to AI projects in his department. His familiarity with the company's operations combined with his new AI skills enabled him to seamlessly move into a full-time AI role.

These stories underscore the varied paths one can take toward a career in AI and tech. They highlight the importance of building on existing skills, continuously learning new ones, and remaining adaptable to navigate this ever-evolving field effectively. Whether you are looking to completely switch careers or aiming to integrate AI into your current role, the key is to start where you are, use what you have, and do what you can. With a structured plan, a learning mindset, and a bit of persistence, transitioning into an AI career is not just a possibility but an exciting opportunity to be at the forefront of technological innovation.

6.5 CONTINUOUS LEARNING: KEEPING UP WITH AI TRENDS

In the ever-evolving field of artificial intelligence, staying abreast of the latest trends and developments is not just beneficial—it's necessary. The pace at which AI evolves can be dizzying, with new technologies, frameworks, and methodologies emerging regularly. This rapid development can render once-cutting-edge knowledge obsolete in just a few years, making continuous learning an integral part of maintaining relevance and expertise in the field. Moreover, the breadth of AI's applications across different industries—from healthcare and finance to entertainment and automotive—means that new use cases and challenges are constantly arising, each requiring a fresh understanding and novel approaches.

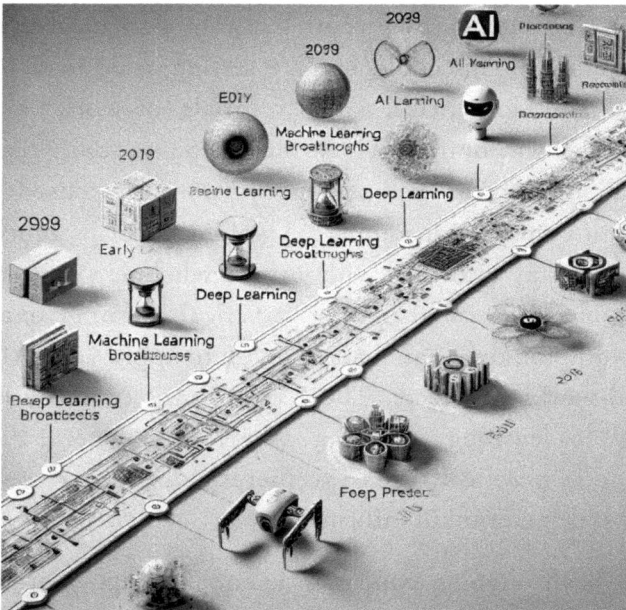

For those keen on keeping up with these developments, a plethora of resources is available. Engaging with reputable AI blogs and newsletters can be a great way to receive curated, up-to-date information on the latest in AI. Websites like 'Towards Data Science' on Medium offer accessible yet insightful articles, often penned by industry experts and academics. Newsletters from major AI research labs such as OpenAI or Google's DeepMind provide updates on groundbreaking research and innovations directly from the forefront of AI development. Additionally, attending AI conferences can be immensely beneficial. Events like the Neural Information Processing Systems Conference (Neur-IPS) or the International Conference on Machine Learning (ICML) present the latest research and provide networking opportunities with leading AI professionals and researchers. These interactions can offer more in-depth insights and foster collaborations that are not typically available through other mediums.

Building a personal learning routine is another crucial strategy for staying updated. This could involve setting aside dedicated time each week to read articles, watch tutorials, or work on personal AI projects. Incorporating AI learning into your daily routine can also be effective. For example, listening to AI-focused podcasts during your commute or joining AI webinars during lunch breaks can help you stay informed without overwhelming your schedule. Websites like Kaggle also offer practical learning experiences through competitions that challenge you to apply your AI skills to solve real-world problems. These platforms provide a hands-on approach to learning, allowing you to implement theoretical knowledge in practical scenarios, thereby deepening your understanding and retention of information.

Engaging with ongoing education is equally important. The field of AI is expanding and deepening, with specializations in areas like natural language processing, computer vision, and robotics. Participating in online courses and workshops can help you delve deeper into these specialties. Platforms such as Coursera and edX collaborate with top universities and companies to offer courses that range from introductory to advanced levels, covering both broad and niche topics within AI. These courses often include peer interaction, practical assignments, and sometimes even real-time sessions with instructors, providing a comprehensive educational experience. For a more immersive learning experience, some professionals opt for boot camps or intensive programs that focus on specific areas of AI. These programs are designed to build expertise quickly and are especially useful for professionals transitioning into AI from other fields.

The journey of learning in AI is continuous and multifaceted. By engaging with a mix of reading, interactive learning, formal education, and community participation, you can develop a robust understanding of both foundational and advanced AI concepts.

This ongoing process of learning enhances your skills and prepares you to adapt to new challenges and opportunities in the field of AI. As AI continues to permeate various aspects of our lives, the knowledge and skills you acquire will enhance your career prospects and equip you with the tools to contribute to the future development of this transformative technology.

6.6 FREELANCING WITH AI SKILLS: FINDING YOUR NICHE

In the expansive world of artificial intelligence, the opportunities for freelancers are growing exponentially. The increasing demand for AI expertise spans numerous industries, from tech startups craving innovative machine-learning solutions to established corporations looking to harness AI for data analysis, customer service improvements, and operational efficiencies. As a freelancer, your ability to offer specialized, scalable AI solutions can set you apart and open doors to diverse project engagements. The key to success lies in identifying your niche — an area within AI that not only interests you but where your skills can solve specific problems for clients.

Identifying your AI freelancing niche requires a strategic approach. Start by assessing your strengths and areas of expertise. Are you adept at creating neural networks, or do you excel in natural language processing? Perhaps your strength lies in predictive analytics or robotics. Once you've pinpointed your strengths, research industries that could benefit most from these capabilities. For instance, if you have a knack for speech recognition technologies, your niche could be developing AI-driven virtual assistants for healthcare providers to manage patient inquiries. By focusing on a niche, you can develop deep expertise that allows you to offer more value to your clients, which is often more lucrative and satisfying than being a generalist.

Marketing your AI freelance services effectively is crucial to attracting the right clients and projects. Develop a clear, compelling value proposition that communicates what you offer and the specific benefits it brings to your clients. Build a professional website showcasing your portfolio of projects and testimonials from past clients. This serves as a tangible demonstration of your skills and successes. Utilize content marketing by writing blog posts or articles that highlight your expertise in your chosen niche. Share these on social media platforms and professional networks like LinkedIn to increase your visibility. Attending industry conferences, either as a participant or a speaker, can also be a powerful marketing tool. These venues offer the opportunity to connect with potential clients and establish yourself as an expert in your field.

Managing a freelance AI business involves more than just technical skills; it requires effective client management, pricing strategies, and maintaining a work-life balance. When setting prices for your services, consider the value you provide, your experience, and market rates. Be transparent with your clients about your pricing model and the expected outcomes of your projects. Effective communication is key to successful client management. Keep your clients updated on project progress and any challenges that arise. This builds trust and can lead to ongoing work and referrals. Finally, managing work-life balance is crucial, especially when different projects might demand varying levels of attention and time. Set clear boundaries for work hours and ensure you allocate time for rest and personal activities. This not only prevents burnout but also keeps you motivated and productive.

Navigating the freelance AI landscape requires a blend of technical proficiency, strategic marketing, and effective business management. By carving out a niche, showcasing your expertise, and managing your projects and clients adeptly, you can build a rewarding and successful freelance career in artificial AI.

As we wrap up this chapter on advancing your career through freelancing with AI skills, remember the importance of continuous learning and adaptation in this dynamic field. The insights and strategies explored here are designed to equip you with the knowledge to navigate the freelancing world confidently, showcasing your unique skills and carving a niche that aligns with your passions and the market's needs. As you move forward, keep honing your skills, expanding your network, and embracing the opportunities that artificial intelligence offers. This journey is not just about professional growth, but also about contributing to a field that is reshaping industries and touching lives around the globe. In the next chapter, we will explore how emerging trends and future predictions in AI continue to offer new opportunities and challenges in this exciting field.

CHAPTER SEVEN

EMERGING TRENDS AND FUTURE
PREDICTIONS

I magine stepping into a studio where the artist is not a person, but an algorithm. Here, paintings, music, and literary works are not born from human hands or minds, but from lines of code. Welcome to the fascinating world of generative AI—the next frontier in artificial intelligence where machines are not just tools, but creators. This rising trend is reshaping the landscape of creativity, offering a glimpse into a future where AI's influence extends beyond practical applications to inspire and innovate in the realms of art and expression.

7.1 GENERATIVE AI: THE NEXT FRONTIER

Exploring generative AI: Introduction to generative AI and its capability to create new content

Generative AI refers to a type of artificial intelligence that is programmed to generate new content, from images and music to text and beyond, based on the data it has been trained on. This technology utilizes advanced algorithms to analyze and learn from a set of data—be it thousands of paintings, hours of music, or vast libraries of text—and then uses this learned information to create new, original works that are often indistinguishable from those created by humans. For instance, generative AI can analyze the styles of classical composers and then compose new symphonies that carry the essence of Beethoven or Mozart but with a completely original score. The capabilities of generative AI are not just reproducing existing styles, but also combining and evolving them to create something entirely new and innovative.

Applications and innovations: Highlighting how generative AI is used in art, music, and text generation

The applications of generative AI are as diverse as they are groundbreaking. In the world of visual arts, AI programs like DeepArt and Artbreeder allow users to create complex images and artworks that mirror the styles of renowned painters or combine different styles to form new, unique creations. In music, tools like OpenAI's Jukebox can compose music in various genres and even simulate the voices of well-known artists, offering possibilities for new forms of musical expression and experimentation. In literature and journalism, AI like GPT-3 is being used to write poems, draft articles, and even create entire novels, pushing the boundaries of how stories are told and conceived.

Ethical considerations: Discussing the ethical implications of AI-generated content and the importance of responsible use

As with any powerful technology, the rise of generative AI brings with it a host of ethical considerations that must be addressed. The ability of AI to replicate and create content that closely mimics human-generated work raises questions about originality, authenticity, and the value of human creativity. There are concerns about copyright and intellectual property rights—issues that become increasingly complex when machines are the creators. Furthermore, the potential use of generative AI in spreading misinformation or creating deepfake content that can be indistinguishable from real images or videos is a significant concern. Developers and users of generative AI must engage in discussions about ethical guidelines and regulatory frameworks to ensure that these technologies are used responsibly and for the benefit of society.

Future potential: Speculating on the untapped potential of generative AI and how it could transform creative industries

Looking ahead, the potential of generative AI to transform creative industries is immense. In the future, we might see AI assisting in the creative process and collaborating with humans to produce hybrid works that blend human intuition with AI's capacity for data-driven creativity. This collaboration could accelerate innovation in fields such as design, where AI could offer new ways to approach complex problems or create sustainable solutions. Additionally, as generative AI becomes more accessible, it could democratize art and creativity, allowing more people to express themselves in ways that were previously limited to those with specific skills or training. The convergence of AI with traditional creative practices is poised to create a new era of artistic expression, characterized by limitless possibilities and an ever-evolving definition of what it means to create.

As we explore these emerging trends and the future potential of AI, it's clear that the technology is poised to not only augment human abilities but also to challenge and expand our understanding of creativity. The journey into the capabilities of genera-tive AI is just beginning, and its trajectory promises to be as exciting as it is unpredictable. Engaging with this technology today may well prepare us for a future where AI and human creativity intermingle seamlessly, creating a world enriched by both technological advancement and artistic expression.

7.2 THE ROLE OF AI IN SUSTAINABLE TECHNOLOGIES

Artificial Intelligence (AI) is significantly altering how we approach sustainability, paving the way for smarter solutions in energy conservation, resource management, and environmental protection. The integration of AI into sustainable practices offers a promising avenue to tackle some of the most pressing environ-mental challenges of our time. By harnessing AI, we can optimize the use of natural resources, enhance energy efficiency, and reduce the ecological footprint of human activities. For instance, AI's ability to analyze large datasets can lead to more informed decisions about energy use, water management, and waste reduc-tion, ensuring that resources are used more efficiently and sustainably.

One of the most impactful applications of AI in sustainability is in the field of renewable energy. AI algorithms help optimize the operation of wind and solar power systems by predicting weather conditions, adjusting the angles of solar panels, or operating wind turbines to maximize energy capture during varying weather conditions. This not only increases the efficiency of renewable energy sources but also makes them more reliable and cost-effective. For example, DeepMind has collaborated with Google to apply machine learning algorithms to 120 wind farms in the U.S. The AI system predicts wind power output 36 hours ahead of actual generation, enabling more effective grid integration and boosting the value of wind energy by roughly 20 percent.

AI's role extends beyond energy production to encompass waste reduction and management. Through sophisticated sorting technologies, AI can significantly enhance recycling processes. Systems equipped with AI-driven sensors and cameras can quickly identify and sort recyclable materials from waste streams, reducing contamination and improving the quality of recycled materials. This not only makes recycling more efficient but also more economically viable. In cities like Shanghai, AI-powered waste sorting and management systems have been implemented to tackle the city's substantial waste management challenges, demonstrating substantial improvements in recycling rates and reductions in landfill use.

However, leveraging AI for sustainability is not without challenges. One of the primary obstacles is the availability and quality of data. For AI systems to function effectively, they require access to accurate and comprehensive data. In many areas of environmental management, such data can be scarce or of poor quality, which can limit the effectiveness of AI applications. Furthermore, there is the issue of the environmental cost of training large AI models. The energy consumption associated with operating massive data centers that train and run AI systems can be substantial, which paradoxically could contribute to the environmental footprint—the very issue AI in sustainability aims to mitigate.

Looking forward, the vision for a sustainable future powered by AI is both optimistic and achievable. Imagine a world where cities are smart and fully integrated with AI technologies that manage everything from traffic and waste to energy and water, all optimized for minimal environmental impact. In agriculture, AI-driven systems could precisely manage the use of water and fertilizers, reducing waste and environmental degradation. Moreover, AI could play a crucial role in biodiversity conservation, monitoring animal populations, and habitats to prevent poaching and encroachment. The potential of AI to support sustainable living extends across all facets of human endeavor, offering hope for a balanced coexistence with our natural environment.

As we explore the capabilities of AI in fostering sustainability, it becomes evident that this technology is not just an auxiliary tool, but a fundamental aspect of crafting a sustainable future. The synergy between AI and sustainability strategies presents a potent solution to ecological challenges, promising a greener, more efficient, and sustainable world for future generations.

7.3 AI IN SPACE EXPLORATION: WHAT LIES AHEAD

The cosmos has always beckoned to humanity, a vast expanse filled with mysteries and the potential for discovery. Today, artificial intelligence (AI) is playing a pivotal role in how we explore this final frontier. The integration of AI into space exploration is not merely an addition; it is transforming the very methodology of how we reach beyond our earthly confines and what we can achieve in the cosmos. AI's role encompasses a range of functions, from enhancing the autonomy of spacecraft to handling vast quantities of space data, each of which contributes significantly to the efficiency and success of missions.

AI's current role in space exploration includes the automation of routine tasks, allowing scientists to focus on more complex problems. For instance, AI algorithms are used to monitor the health of spacecraft, analyze the telemetry data, and even make decisions about the best course of action in real time. This kind of autonomy is crucial, especially for long-duration space missions where direct human oversight is impossible due to the vast distances involved. Moreover, AI is instrumental in processing the immense volumes of data that space missions generate. It helps identify patterns and insights that might be overlooked by human analysts, such as subtle changes in planetary atmospheres or irregularities in asteroid surfaces. This capability not only enhances our understanding of space phenomena but also informs future missions and strategies.

The potential for unmanned missions with AI-driven robots and spacecraft is particularly exciting. These missions could explore the surfaces of distant planets or moons, conducting experiments and collecting samples, without the risk to human life. Imagine a rover, driven not by direct commands from Earth, but by an AI system capable of navigating alien terrains autonomously. Such rovers could make real-time decisions about where to go, what to study, and even how to react to unexpected situations. This would increase the efficiency of space exploration and expand the scope of what can be achieved. For instance, AI-driven probes could venture into environments that are too hostile for humans, such as the icy oceans believed to be beneath the surface of Jupiter's moon Europa, potentially discovering signs of life.

Data analysis and discovery through AI also hold transformative potential for space exploration. Space missions generate an enormous amount of data, from high-resolution images of distant galaxies to intricate measurements of cosmic radiation. AI excels in managing and interpreting this data, uncovering patterns and correlations that can lead to new scientific discoveries. For example, AI algorithms have been used to sift through data from telescopes to identify new exoplanets—planets beyond our solar system. These algorithms can detect the minute dimming of stars caused by planets passing in front of them, a task that is challenging and time-consuming for human researchers. By rapidly processing data, AI not only accelerates the pace of discovery but also enhances the depth and breadth of scientific research in astronomy.

Looking toward the future, the collaboration between AI and humans in space exploration is likely to deepen, creating new paradigms for how we explore and interact with space. The development of AI systems that can work alongside humans, assisting with tasks such as habitat construction or life support management, is underway. These systems would need to handle technical tasks and adapt to the nuances of human behavior and needs, creating a symbiotic relationship between human astronauts and AI. For instance, during long-duration missions to Mars, AI could manage life support systems, optimize resource use, and even provide psychological support for astronauts, helping to maintain crew health and morale. This human-Ai collaboration could extend to the colonization of other planets, where AI systems might oversee the construction of habitats or the management of agricultural ecosystems, essential for sustaining human colonies.

As we continue to push the boundaries of what is possible in space exploration, AI stands as a crucial ally. From enhancing the autonomy of spacecraft to handling interstellar data, AI's role is not only supportive but foundational in the quest to explore the cosmos. The future of space exploration, with AI as a central component, promises not only more efficient and ambitious missions but also a more in-depth understanding of the universe we inhabit. The potential for AI to transform space exploration is as boundless as space itself, heralding a new era of discovery and adventure beyond the stars.

7.4 THE FUTURE OF AI IN EDUCATION

Imagine a classroom that adapts to the pace and style of every student, creating a truly personalized learning environment. This is not a distant reality but a potential present with the integration of Artificial Intelligence (AI) in education. AI can revolutionize how educational content is delivered, making learning a more tailored and accessible experience. Personalized learning through AI involves algorithms that analyze a student's performance on various tasks and adjust the curriculum accordingly. For example, if a student excels in mathematical reasoning but struggles with data interpretation, the AI system can modify subsequent content to reinforce data skills while continuing to challenge areas of strength.

This personalization extends beyond academic skills to adapt to individual learning preferences and needs. Some students might benefit from visual representations of information, while others might prefer auditory explanations or hands-on activities. AI systems can identify these preferences and deliver content in formats that maximize an individual's learning potential. Additionally, these systems can monitor a student's engagement and fatigue levels, suggesting breaks or changes in activity when needed. This responsive approach can help maintain optimal levels of engagement, crucial for effective learning.

The potential of AI doesn't stop at personalized learning; it extends to providing support through AI tutors and assistants. These AI-driven tools offer students on-demand help with homework, clarification of complex topics, and additional practice in areas of difficulty, all without the need for human intervention. For instance, an AI tutor could help a student solve a complicated algebra problem step-by-step, providing explanations for each step and adjusting the difficulty based on the student's responses.

This kind of support is particularly valuable in large classrooms, where teachers may not always have the time to address every individual student's needs. Moreover, AI tutors are available around the clock, assisting outside of school hours, thus extending learning opportunities.

However, the integration of AI in education is not without its challenges. Balancing the benefits of AI with concerns about data privacy and the role of teachers is paramount. As AI systems require access to students' performance data to function effectively, ensuring the privacy and security of this data is a significant concern. There must be stringent measures to protect student information from unauthorized access and to ensure that it is used solely for educational purposes. Furthermore, the role of teachers remains irreplaceable and crucial. AI may change the nature of teaching by offloading some instructional duties to AI systems, but teachers will always be needed to provide expertise, context, and emotional support that AI cannot offer. The future lies in finding the right balance, where AI enhances the educational experience without supplanting the human touch that is essential to effective learning.

Envisioning an educational system enhanced by AI, we see a model where AI and human educators work in tandem to optimize learning outcomes. In such a system, AI takes on the role of personalizing learning and providing additional support, while teachers focus on curriculum design, mentoring, and addressing the social and emotional needs of their students. This collaborative approach could lead to more effective education strategies, where technology and human insight combine to prepare students not just academically but as well-rounded individuals ready to tackle the challenges of the future.

As we look toward this future, it becomes clear that AI has the potential to transform the educational landscape dramatically. From personalized learning environments to AI-driven tutoring, the possibilities are vast and promising. However, as we navigate this transformation, it is crucial to keep in mind the ethical implications and the indispensable role of human educators. By steering this technology with a focus on enhancing educational outcomes and maintaining the human element, we can ensure that AI serves as a powerful tool for enriching education and empowering learners.

7.5 ETHICAL AI: SHAPING A RESPONSIBLE FUTURE

In the rapidly evolving landscape of artificial intelligence, the conversation around ethics becomes not just relevant, but imperative. As AI technologies assume roles from automating mundane tasks to making complex decisions, the need to embed ethical considerations at the core of AI development and use cannot be overstated. Ethical AI refers to the practice of creating AI technologies that comply with legal standards and adhere to widely accepted ethical norms and principles. These principles, such as fairness, accountability, and transparency, guide the responsible development and deployment of AI systems.

At the heart of ethical AI lies the commitment to fairness. This principle ensures that AI systems do not create or perpetuate discrimination and are equitable in their operations. For instance, an AI hiring tool should be designed to assess candidates based on relevant qualifications without bias towards gender, race, or age. Achieving this requires rigorous testing and training of AI systems on diverse datasets to minimize any inherent biases that may exist. Moreover, fairness extends beyond preventing bias; it also involves ensuring that AI technologies are accessible to all, thereby democratizing the benefits of AI across different sectors of society.

Transparency and accountability are equally critical in the ethical use of AI. Transparency in AI implies that the workings of AI systems are understandable to users and other stakeholders. This is particularly important in applications where AI decisions have significant impacts on individuals' lives, such as in healthcare or criminal justice. For AI systems to be transparent, they should not operate as 'black boxes'. There should be clarity on how decisions are made, and the factors influencing these decisions should be openly communicated. This openness builds trust and allows for accountability, where developers and users of AI systems are held responsible for the outcomes of AI decisions. Ensuring accountability involves setting up mechanisms for feedback and redress, allowing individuals to challenge unfair or incorrect decisions made by AI systems.

Global initiatives and frameworks have been pivotal in shaping the standards for ethical AI. These efforts aim to create a unified approach to govern AI ethics globally. One notable framework is the guidelines set by the European Union, which focus on trustworthy AI. These guidelines emphasize AI systems being lawful, ethical, and robust, from their inception through their operational life. Similarly, the OECD Principles on AI provide a global standard that aligns with values of fairness, transparency, and accountability, endorsed by over 40 countries. These frameworks are not just theoretical; they influence policies and practices that govern AI applications worldwide, ensuring that AI advances with consideration for its broader impacts on society.

Looking ahead, the landscape of AI ethics is poised to encounter new challenges. As AI technologies become more complex and ingrained in our daily lives, the ethical dilemmas we face will evolve. Questions about the extent to which AI should make decisions in human affairs, the privacy implications of AI in surveillance, and the use of AI in warfare are just some of the ethical quandaries that will require rigorous debate and thoughtful solutions. Moreover, as AI continues to transcend international borders, the challenge of implementing cohesive global standards for AI ethics becomes more pronounced. Different cultural and social norms may influence perceptions of what is considered ethical, making universal agreements challenging.

Navigating the future of AI ethics will involve a collaborative effort among governments, technology developers, civil society, and the public. It will require a commitment to continuous learning and adaptation, as the technologies and their societal impacts evolve. Engaging in open dialogues, fostering education on AI ethics, and encouraging inclusive participation in these discussions will be crucial. As we advance further into the age of artificial intelligence, our ability to integrate ethical considerations into the fabric of AI development will significantly determine the trajectory of its impact on humanity. The path we forge today in AI ethics will lay the foundation for a future where AI not only enhances our capabilities but does so in a way that aligns with our shared values and aspirations.

7.6 THE CONVERGENCE OF AI AND VIRTUAL REALITY

Imagine stepping into a world where your physical limitations don't define your capabilities—a world where training, learning, and entertainment transcend conventional boundaries. This is the evolving reality of virtual reality (VR) enhanced by artificial intelligence (AI). When AI meets VR, the experience shifts from mere visual immersion to an interactive environment that responds intelligently to your actions, enhancing both the realism and the personal relevance of the experience.

The synergy between AI and VR brings a multitude of enhancements to virtual environments, making them more dynamic and intuitive. AI contributes to VR by improving the environment's responsiveness through adaptive algorithms that learn from your interactions. For example, in a virtual reality game, AI can analyze how you play and adjust the game's difficulty in real-time, ensuring a consistently challenging and engaging experience. Furthermore, AI can populate VR worlds with intelligent characters that react to your actions in realistic ways, enhancing the immersive experience. These characters can lead conversations, adapt their behavior based on your responses, and provide a level of interaction that feels astonishingly real.

In professional settings, the integration of AI with VR is transforming training and simulation. Industries like healthcare, aviation, and manufacturing are already reaping the benefits of this technology. In medical training, for instance, VR simulations powered by AI can mimic complex surgical procedures, providing trainees with a realistic and interactive environment to hone their skills without any risk to real patients. The AI component allows the simulation to react to the trainee's actions, providing instant feedback that is crucial for learning and improvement. Similarly, in aviation, pilots can use VR simulations to train under various flight conditions. AI enhances these simulations by introducing unexpected changes in weather or technical issues, preparing pilots for real-world scenarios that they might not otherwise experience until they face them in flight.

The social and entertainment aspects of AI-driven VR are equally promising. Social VR platforms allow people to interact in a virtual space that mimics real-world interactions, offering a new dimension to online socializing. AI enhances these interactions by analyzing user behavior to improve avatar expressions and movements, making the interaction more lifelike. Additionally, in the entertainment sector, AI-driven VR can tailor experiences to individual tastes and preferences, much like the algorithms used by streaming services. For example, a VR concert experience could adapt the set list in real-time based on the crowd's reactions, picked up and analyzed by AI.

Anticipating future developments, the integration of AI and VR is set to deepen, driven by advancements in both fields. As AI becomes more sophisticated, its integration into VR will offer increasingly seamless and responsive experiences. Future VR could include more nuanced emotional recognition, allowing AI to respond not just to what users do, but how they feel, by analyzing voice tones and body language. This could lead to applications in mental health, where VR environments soothe or stimulate users based on their emotional state, monitored and managed by AI.

The convergence of AI and VR is not just about technological advancement; it's about creating a fabric of experiences that are profoundly personalized and immensely enriching. As these technologies evolve, they promise to redefine our perception of reality and expand our possibilities for interaction and exploration. This exciting confluence is not merely about escaping reality but enhancing it, opening new avenues for growth, learning, and entertainment that were once the domain of science fiction.

In conclusion, as we explore the vast potentials of AI-enhanced virtual reality, we see a future where our digital interactions are as complex and meaningful as our interactions in the physical world. The convergence of AI and VR holds the promise of transforming not only how we entertain and educate ourselves, but also how we connect with each other in a deeply immersive, interactive digital environment. This chapter sets the stage for understanding how intertwined technologies continue to shape a future where the virtual and real increasingly reflect and enhance one on another, leading us into the next chapter of our exploration of AI's impact across different facets of life and society.

CHAPTER EIGHT

BUILDING AN AI-INCLUSIVE WORLD

I magine a world where the power of artificial intelligence (AI) is not only a catalyst for innovation and efficiency but also a potent tool for social good. This world is not far-fetched. Today, AI is increasingly playing a pivotal role in addressing complex social challenges, offering solutions that were once thought impossible. As we explore this transformative journey, you'll discover how AI is not just about technology; it's about the profound impact it can have on society, improving lives and empowering communities globally.

8.1 AI FOR SOCIAL GOOD: CASE STUDIES AND OPPORTUNITIES

Highlighting Successful AI Projects: Presenting Case Studies Where AI Has Been Used to Address Social Issues and Improve Lives

The potential of AI to drive social change is immense, as illustrated by several groundbreaking projects. One such initiative is the "AI for Social Good" program launched by Google. This initiative harnesses AI to tackle some of the world's most pressing challenges, such as environmental conservation and social welfare. For instance, Google's AI has been used to analyze weather patterns and predict flooding, significantly improving disaster response and potentially saving thousands of lives. Another inspiring example is the use of AI in healthcare, particularly in developing countries. AI-driven applications have been developed to diagnose diseases such as tuberculosis and diabetes more quickly and accurately, greatly enhancing treatment outcomes for communities with limited access to medical specialists.

These case studies underscore AI's potential to not only transform industries but also to create a tangible positive impact on society. They serve as powerful examples for you, whether you're an AI enthusiast, a developer, or simply someone interested in the intersection of technology and social good. By examining these projects, you can gain insights into how AI technologies are applied in real-world scenarios to solve complex problems and improve human well-being.

Opportunities for Impact: Identifying Areas Where AI Has the Potential to Make Significant Positive Social Impacts

The scope for AI to contribute to social good extends across various domains, from education and healthcare to environmental sustainability and humanitarian aid. In education, AI can personalize learning, adapting educational content to meet the unique needs of each student, thus democratizing access to quality education. In environmental conservation, AI technologies are used to monitor wildlife and track poaching activities, helping preserve biodiversity. Moreover, AI can optimize energy consumption in cities, contributing to more sustainable urban development.

For you, understanding these opportunities means recognizing the broad spectrum of applications where AI can be leveraged for social good. It opens up avenues for innovation and involvement, whether by supporting AI initiatives or by initiating projects in your community. The potential for AI to drive social change is limited only by our imagination and commitment to applying these technologies ethically and effectively.

Overcoming Barriers to Access: Strategies for Ensuring that the Benefits of AI Are Widely Accessible, Regardless of Socioeconomic Status

Despite its potential, the benefits of AI are not yet accessible to everyone. A significant barrier is the digital divide that limits access to AI technologies in underprivileged areas. Overcoming this requires concerted efforts to increase connectivity, improve digital literacy, and ensure that AI solutions are designed to be inclusive and accessible to all segments of society.

Initiatives such as AI4ALL seek to address these challenges by educating and empowering the next generation of AI users and developers from diverse backgrounds. These programs focus on inclusive education and community engagement to ensure that the benefits of AI extend to all sectors of society, particularly those who are most in need of its transformative power.

Encouraging Participation: Inspiring Readers to Get Involved in AI for Social Good Projects, Either as Developers or as Advocates

Getting involved in AI for social good can take many forms, from developing new solutions to advocating for ethical AI practices. You can participate in hackathons focused on social good, contribute to open-source projects that are making a difference, or support policies that promote the ethical use of AI. Each action contributes to a larger movement towards a more equitable and sustainable future powered by AI.

For developers, this might mean using your skills to create applications that solve specific social issues. For non-developers, involvement could be in the form of supporting organizations that are working on AI for social good, or simply staying informed and spreading the word about the positive potential of AI.

As AI continues to evolve, the opportunities for it to support social good initiatives grow. By engaging with AI, whether through direct development or advocacy, you contribute to a movement that uses technology as a force for good, shaping a future where AI enhances everyone's quality of life, not just the privileged few. This is your call to action—an invitation to be part of a transformative process that leverages cutting-edge technology for the betterment of society.

8.2 BRIDGING THE DIGITAL DIVIDE WITH AI EDUCATION

Understanding artificial intelligence (AI) is becoming increasingly important for everyone, not just those in tech-centric careers. As AI technologies continue to shape various sectors—from healthcare to finance—it is crucial that everyone has a basic understanding of AI, ensuring they can navigate this new landscape effectively and make informed decisions. AI literacy is not just about understanding the technical workings of algorithms but also about recognizing their impact on our daily lives, privacy, job security, and ethical considerations. It empowers individuals to advocate for responsible AI use and understand potential biases in AI applications. As AI continues to evolve, the divide between those who understand AI and those who do not could widen, leading to disparities in how individuals and communities interact with, benefit from, or are harmed by AI technologies.

To address this, numerous initiatives have been launched worldwide to promote AI education, particularly targeting underserved communities. For instance, programs like AI4ALL open doors for high school students from underrepresented backgrounds to learn about AI through hands-on projects and mentorship, aiming to create a diverse pipeline of AI talent. Universities and online platforms are also expanding their curriculum to offer AI courses that cater to a broader audience, including courses that require no prior programming knowledge. These educational initiatives are crucial in democratizing AI knowledge, making it accessible to a wider audience, and ensuring that the future of AI is shaped by a diverse group of thinkers and makers.

However, numerous challenges hinder the goal of universal AI education. Access to quality education resources remains a significant barrier. In many areas around the world, students and educators lack the necessary tools such as reliable internet access and modern computing devices to engage with digital education platforms. Furthermore, there is often a lack of qualified instructors who can demystify AI concepts and teach them in a relatable way. Overcoming these obstacles requires concerted efforts from governments, educational institutions, and private organizations. Investment in digital infrastructure, teacher training, and curriculum development must be prioritized to ensure that AI education is not a privilege for the few but a fundamental right accessible to all.

Now, let me encourage you, whether you're a student, educator, policymaker, or simply an interested individual, to take action towards making AI education more inclusive. Engage with local schools and community centers to introduce AI topics, advocate for the inclusion of AI education in school curricula, or support non-profit organizations that are working towards this goal. Every effort counts in bridging the digital divide and ensuring that the benefits of AI do not just enrich a select few but enhance the lives of many. Your involvement can make a crucial difference in preparing society for a future where AI plays a central role.

8.3 THE GLOBAL IMPACT OF AI ON EMPLOYMENT

As artificial intelligence (AI) continues to evolve and integrate into various sectors, its impact on the job market is a topic of robust discussion and analysis. Across industries, from manufacturing to financial services, AI is reshaping the nature of work. This transformation presents a dual-edged sword: while AI can streamline processes and enhance efficiency, it also poses challenges in terms of job displacement. Understanding this dynamic is crucial for preparing a workforce that can thrive in an AI-driven economy.

AI's influence on employment is multifaceted. In sectors like automotive manufacturing, robots have been employed for years to perform repetitive tasks more efficiently than humans. However, AI's capability to analyze large datasets and automate decision-making processes is pushing its application into areas previously thought immune to automation, such as certain aspects of legal work and even journalism. These changes suggest a shift in the types of jobs that will be available, with a decrease in demand for roles that involve routine tasks and an increase in roles requiring complex problem-solving skills and creativity.

The transformation is not just about loss but about evolution. New job categories are emerging as AI creates opportunities for employment in AI development, supervision, and maintenance. For instance, as AI systems become more prevalent, the demand for AI safety engineers who can ensure these systems operate as intended is expected to rise. Similarly, there is an increasing need for data scientists and analysts who can interpret the vast amounts of data generated by AI systems. This shift highlights the importance of adaptability in the workforce.

Preparing for these changes involves significant reskilling and upskilling efforts. Workers need to be equipped with new skills to match the evolving demands of the labor market. Educational institutions and businesses must play pivotal roles in this transition by providing learning opportunities that are aligned with the changing job landscape. Programs that focus on digital literacy, coding, AI ethics, and data analysis will be particularly beneficial. Additionally, lifelong learning must become a norm, with individuals taking proactive steps to continually update their skills throughout their careers.

The role of policymakers in this transition is critical. Governments have the responsibility to create policies that encourage and facilitate workforce transformation, while also providing safety nets for those displaced by AI. This includes funding for education programs, incentives for businesses that invest in employee training, and support for innovation in AI governance. Policies need to strike a balance between promoting technological advancement and protecting workers, ensuring that the benefits of AI are distributed broadly across society.

In discussing AI's impact on employment, it's essential to maintain a balanced perspective. While it's true that AI can automate jobs, leading to displacement, it also holds the potential to create new opportunities that could lead to a more prosperous and innovative society. The key lies in proactive preparation and inclusive policies that ensure all members of society have the tools to adapt and thrive in this new era. By fostering an environment that encourages continual learning and adaptation, we can leverage AI not as a force for disruption but as a catalyst for empowering a dynamic, skilled workforce capable of tackling the challenges of the future.

8.4 ADVOCATING FOR DIVERSITY IN AI DEVELOPMENT

When we discuss the development of artificial intelligence, the conversation inevitably turns towards the need for diverse perspectives. It's not just a matter of fairness or representation; diversity in AI development directly influences the effectiveness and inclusiveness of AI solutions. Imagine an AI system trained to recognize voices but developed primarily by and with data from male voices. Such a system would likely underperform when interacting with female voices, not due to any inherent limitation in the technology but because of the homogeneous nature of its training data and development team. This is a simple illustration of how a lack of diversity can skew AI performance and impact. By incorporating a broad range of perspectives—across gender, race, ethnicity, and beyond—AI technologies can be better equipped to serve a wider segment of society, reflecting the rich tapestry of human experience and needs.

However, the path to achieving diversity in AI is fraught with challenges. The tech industry, and AI development in particular, has traditionally been dominated by a relatively narrow demographic. This lack of diversity is not just in gender and ethnicity but also in socioeconomic background and educational pathways. The implications are significant: AI systems are being built that may not fully understand or appropriately respond to the needs of diverse populations. For instance, facial recognition technologies have been shown to have lower accuracy rates for women and people of color, a direct consequence of the lack of diverse data and developers in the training phases of these systems.

To combat these disparities, numerous initiatives have emerged aimed at increasing representation in AI development. Organizations such as Black in AI and Women in Machine Learning are working to foster inclusion and provide support for underrepresented groups in the AI field. These organizations not only advocate for diversity but also create opportunities for networking, mentorship, and career advancement. Furthermore, some tech companies are beginning to recognize the importance of diversity and are instituting programs to recruit and retain more diverse talent. These efforts are crucial, as they help to broaden the pool of perspectives that shape AI technologies, ensuring these tools are more reflective of and responsive to the needs of a diverse global population.

Looking forward, the commitment to promoting diversity in AI development must be sustained and intensified. This means not only continuing to support and expand the reach of existing diversity programs but also embedding diversity considerations into the core practices of AI research and development. Companies and research institutions can start by auditing their AI systems for bias regularly and openly sharing the outcomes, thereby holding themselves accountable. Educational programs in AI should also incorporate discussions about the ethical implications of biased AI systems and train students to think critically about how and why AI systems should be developed responsibly.

For individuals, becoming an advocate for diversity in AI can take many forms. It might involve supporting policies that promote inclusiveness in technology, participating in or donating to organizations that foster diversity in AI, or simply educating oneself and others about the importance of this issue. Each action contributes to a broader movement towards a more equitable AI future.

In the realm of artificial intelligence, diversity is not an optional add-on—it is a fundamental component that ensures the technologies we develop are as fair, effective, and inclusive as possible. As we continue to innovate and push the boundaries of what AI can do, let us ensure that these technologies are shaped by a diverse set of minds and hands. Only then can we fully realize the trans-formative potential of AI to benefit all of humanity.

8.5 AI ACCESSIBILITY: MAKING TECH USABLE FOR EVERYONE

Understanding what it means for AI technologies to be accessible to people with disabilities is a crucial step toward creating a world where everyone can benefit from the advances of technology. Accessibility in AI refers to the design of AI systems that are usable by people with a wide range of abilities and disabilities. This encompasses the development of AI applications that can be effectively used by people with visual, auditory, physical, speech, cognitive, and neurological disabilities. For instance, voice-activated AI helps individuals with visual impairments to interact with technology through spoken commands, while text-to-speech technologies provide auditory access to content that would otherwise be inaccessible.

Currently, the state of AI accessibility varies significantly across different technologies and applications. While some areas have seen notable advances, such as the development of AI-powered assistive devices that help visually impaired individuals navigate their environments, other areas lag behind. A major challenge is ensuring that AI systems are designed from the ground up with accessibility in mind, rather than retrofitting solutions to existing technologies. Many AI tools and applications are not yet fully accessible, often because accessibility considerations were not integrated into the early stages of design and development. This oversight can lead to the exclusion of individuals with disabilities from using these technologies, or from benefiting fully from what AI has to offer.

Adopting best practices for accessible AI design is essential for developing technology that enhances the lives of all users, including those with disabilities. These practices include involving people with disabilities in the design process, adhering to established accessibility guidelines, and continuously testing AI systems with diverse groups to identify and address accessibility barriers. For example, ensuring that AI interfaces can interact seamlessly with screen readers and other assistive technologies can make a significant difference in how accessible an AI application is. Moreover, providing alternative ways to interact with AI, such as through voice commands or gesture-based inputs, can help accommodate users with different types of disabilities.

A compelling illustration of successful AI accessibility is seen in AI-powered apps designed to translate sign language into text or speech in real-time, enabling better communication for individuals who are deaf or hard of hearing. Another example is AI systems that customize learning experiences to suit the needs of students with learning disabilities, such as dyslexia, by adapting content presentation and pacing to optimize learning. These case studies not only demonstrate the potential of AI to improve accessibility but also highlight the positive impacts of inclusive technology design. By learning from these successes and continuing to innovate in accessible AI design, developers can create more inclusive technologies that allow everyone to participate fully in a digitally driven world.

As AI continues to permeate various aspects of life, the importance of making these technologies accessible to everyone cannot be overstated. By embracing best practices in AI accessibility, engaging diverse groups in the development process, and continually striving to eliminate barriers, we can ensure that AI serves as a tool for empowerment and inclusion. This approach not only enriches the lives of individuals with disabilities but also enhances the overall quality and utility of AI systems. As we look to the future, let's commit to a vision of AI that embodies the principles of accessibility and inclusion, paving the way for a more equitable technological landscape.

8.6 ENVISIONING AN AI-EMPOWERED SOCIETY: OUR ROLE AND RESPONSIBILITY

As we stand on the brink of a transformative era shaped by artificial intelligence, it's crucial to reflect on the broad societal benefits that responsible AI development and use can bring. AI, when developed with a conscientious mindset, holds the power to revolutionize industries, streamline government operations, enhance the efficiency of humanitarian aid, and significantly improve the quality of life for people around the globe. For example, AI-driven analytics can optimize energy usage across cities, reducing waste and promoting sustainability. In healthcare, AI can predict outbreaks of diseases, allowing for quicker and more targeted responses, ultimately saving more lives.

However, the path to a truly beneficial AI-empowered society is lined with ethical considerations. As we integrate more AI systems into the fabric of daily life, we must ensure these systems are developed with the highest ethical standards. This includes transparency in how AI algorithms make decisions, fairness in AI outcomes regardless of one's background, and accountability for the impacts these technologies have on individuals and communities. Establishing robust ethical guidelines and practices isn't just about preventing misuse; it's about actively shaping AI as a force for good, ensuring it contributes positively to society. Reflecting on these ethics encourages developers and users alike to consider the wider implications of AI technology, from privacy concerns to socio-economic impacts.

The role that each individual and community plays in guiding the development of AI cannot be overstated. This is not a passive journey. Whether you are a software developer, a policymaker, or just someone interested in the future of technology, your voice matters. Engaging with AI technology—whether by learning about it, developing it, or discussing its societal implications—helps ensure that its evolution is aligned with the broader needs and values of society. Community-driven initiatives, public discourse on AI policy, and grassroots advocacy play integral roles in democratizing AI, making sure it serves the many rather than the few.

Looking forward with a hopeful outlook, it's clear that our collective action today will define the AI of tomorrow. Encouraging widespread participation in AI development and governance helps cultivate a diverse array of perspectives and ideas, enriching the AI systems of the future. Imagine a world where AI not only powers machines but also empowers human creativity and ingenuity. This future is not only possible; it is within our reach if we choose to take proactive steps towards it. Let this be a call to action for all of us to engage with AI in a way that promotes a more inclusive, ethical, and prosperous future for everyone.

In this chapter, we've explored the transformative potential of AI and the critical role ethics play in its development. We've also discussed how individual and collective action is essential in steering AI towards positive societal impacts. As we move forward, let's carry these insights into our continued exploration of AI's capabilities and challenges. Let's ensure that as we develop these advanced technologies, we remain steadfast in our commitment to an AI-empowered society that upholds our shared values and works for the common good. In the next chapter, we will delve deeper into the innovative realms AI is touching, exploring cutting-edge advancements and the exciting possibilities they hold.

CONCLUSION

As we reach the conclusion of this enlightening journey through the world of artificial intelligence, it's important to reflect on the ground we've covered together. From the basic principles and misconceptions of AI to the cutting-edge applications that are shaping our world, we've traveled a path designed to demystify AI and make it accessible and engaging for everyone. Whether you started this book with a sense of curiosity or a bit of apprehension about AI, I hope you now feel more confident and informed about how these technologies work and their potential impact on our future.

Throughout this book, our core objectives have been clear and steadfast: to unveil the mysteries of AI, to present its practical, real-world applications, and to delve into the ethical considerations that accompany its development. We've explored foundational concepts like machine learning and neural networks, introduced practical tools and projects for hands-on experience, and discussed the significant influence of AI on career paths and the global job market. Moreover, we've ventured into the future implications of AI, examining emerging trends and the ongoing ethical debates that shape this dynamic field.

One of the most empowering outcomes of gaining knowledge is the confidence it instills. Understanding AI doesn't just prepare you for changes; it enables you to actively participate in these changes. The insights gained from this book equip you to navigate the future more adeptly, fostering a readiness to adapt and innovate as technology continues to evolve.

Now, as we close this chapter, I encourage you not to see it as the end of your learning journey but as a springboard into a continuous exploration of artificial intelligence. The field of AI is ever-evolving, and staying updated with the latest developments is crucial. Engage with new studies, participate in online forums, and apply your knowledge in various aspects of your life and work. Your proactive engagement will not only enhance your personal and professional growth but also contribute to the responsible development and use of AI technologies.

Remember, the future of AI is not just in the hands of scientists and tech experts; it's also in yours. Every individual has a role to play in shaping how this technology evolves and is implemented. By keeping ethical considerations at the forefront, you can help ensure that AI develops in a way that is inclusive, fair, and beneficial for all. Let's strive to be not just consumers of AI technology but active participants in its narrative, contributing to a world where AI enhances our capabilities and enriches our lives.

I am optimistic about the possibilities that AI holds for solving complex global challenges and improving everyday life. Envision a future where AI not only automates tasks but also amplifies our creative and intellectual strengths, leading to breakthroughs in healthcare, environmental protection, and beyond.

Thank you for joining me on this journey of discovery. Your curiosity and engagement have been instrumental in exploring the vast potential of AI. I encourage you to share your newfound knowledge and enthusiasm with others. Together, let's spread the vision of a technologically empowered society, ready to embrace the opportunities that AI brings. Remember, the future is not just something we enter; it's something we create. Let's build it together with knowledge, creativity, and ethical foresight.

REFERENCES

Choi, J. H. (2017). The history of artificial intelligence. *Science in the News*. https://sitn.hms.harvard.edu/flash/2017/history-artificial-intelligence/

University of Oxford. (2024, January 3). Study shows that the way the brain learns is different from the way artificial intelligence systems learn. *University of Oxford News*. https://www.ox.ac.uk/news/2024-01-03-study-shows-way-brain-learns-different-way-artificial-intelligence-systems-learn

Analytics Emerging India. (2020, April 12). Revolutionizing industries with deep learning: Real-world applications and success stories. *Medium*. https://medium.com/@analyticsemergingindia/revolutionizing-industries-with-deep-learning-real-world-applications-and-success-stories-e2235d172926

Gura, T. (2020, October 13). Ethical concerns mount as AI takes bigger decision-making role. *Harvard Gazette*. https://news.harvard.edu/gazette/story/2020/10/ethical-concerns-mount-as-ai-takes-bigger-decision-making-role/

StoryLab AI. (n.d.). Scale up your social media personalization with these 5 AI tools. *StoryLab AI*. https://storylab.ai/ai-for-personalizing-social-media-content-at-scale/

Forbes Technology Council. (2024, January 26). Navigating the AI era: The rise of personalized, real-time e-commerce experiences. *Forbes*. https://www.forbes.com/sites/forbestechcouncil/2024/01/26/navigating-the-ai-era-the-rise-of-personalized-real-time-e-commerce-experiences/

Advanced. (n.d.). Natural language processing (NLP): The science behind chatbots and voice assistants. *Advanced*. https://www.oneadvanced.com/news-and-opinion/natural-language-processing-nlp-the-science-behind-chatbots-and-voice-assistants/

Jiang, F., Jiang, Y., Zhi, H., Dong, Y., Li, H., Ma, S., Wang, Y., Dong, Q., Shen, H., & Wang, Y. (2017). Artificial intelligence in healthcare: Past, present and future. *Stroke and Vascular Neurology*, *2*(4), 230–243. https://doi.org/10.1136/svn-2017-000101

Edureka. (n.d.). Artificial intelligence algorithms for beginners. *Edureka*. https://www.edureka.co/blog/artificial-intelligence-algorithms/

Iberdrola. (n.d.). Data mining: Definition, examples and applications. *Iberdrola*. https://www.iberdrola.com/innovation/data-mining-definition-examples-and-applications

DATAVERSITY. (2023, January 5). AI and big data: How artificial intelligence is transforming the business landscape. *DATAVERSITY*. https://www.dataversity.

net/ai-and-big-data-how-artificial-intelligence-is-transforming-the-business-landscape/

West, S. M., Whittaker, M., & Crawford, K. (2019). Algorithmic bias detection and mitigation: Best practices and policies to reduce consumer harms. *Brookings*. https://www.brookings.edu/articles/algorithmic-bias-detection-and-mitigation-best-practices-and-policies-to-reduce-consumer-harms/

Botpress. (2024). 9 best AI chatbot platforms: A comprehensive guide (2024). *Botpress*. https://botpress.com/blog/9-best-ai-chatbot-platforms

ProjectPro. (2024). 20 artificial intelligence project ideas for beginners [2024]. *ProjectPro*. https://www.projectpro.io/article/artificial-intelligence-project-ideas/461

DigitalOcean. (2024). 10 AI tools transforming web development in 2024. *DigitalOcean*. https://www.digitalocean.com/resources/article/ai-tools-web-development

UST. (2023). Navigating the ethical landscape of AI content creation. *UST*. https://www.ust.com/en/insights/navigating-the-ethical-landscape-of-ai-content-creation

Google. (n.d.). Learn AI skills with Google AI Essentials. *Grow with Google*. https://grow.google/ai-essentials/

Lozovsky, A. (2023). A beginner's guide to the OpenAI API: Hands-on tutorial. *DataCamp*. https://www.datacamp.com/tutorial/guide-to-openai-api-on-tutorial-best-practices

Jaiswal, E. (2023, November 12). Top trending AI tools of 2023 you must know. *Medium*. https://medium.com/@emmaja/top-trending-ai-tools-of-2023-you-must-know-0c840754b3ca

Lozovsky, A. (2023). Building your first AI model: A beginner's step-by-step guide. *LinkedIn*. https://www.linkedin.com/pulse/building-your-first-ai-model-beginners-step-by-step-lozovsky-mba-tmeac

MyGreatLearning. (2023). What are the most in-demand skills in artificial intelligence? *MyGreatLearning*. https://www.mygreatlearning.com/blog/most-in-demand-skills-in-artificial-intelligence/

LinkedIn. (2023). How to showcase AI projects on your resume. *LinkedIn*. https://www.linkedin.com/advice/0/what-best-ways-showcase-ai-projects-your-94xbe

NadiaSpeaks. (2023). 5 networking tips in an AI world! *NadiaSpeaks*. https://nadiaspeaks.com/5-networking-tips-in-an-ai-world/

Maharaj, S. (2023, November 23). How to successfully make a career switch into AI and get a high-paying job. *Medium*. https://medium.com/@sahirmaharaj/how-to-successfully-make-a-career-switch-into-ai-and-get-a-high-paying-job-5c0a9e5f6dfc

McKinsey & Company. (2023). The state of AI in 2023: Generative AI's breakout year. *McKinsey & Company*. https://www.mckinsey.com/capabilities/quantumblack/our-insights/the-state-of-ai-in-2023-generative-ais-breakout-year

Yes Energy. (2023). Using ML and AI in the energy sector (with case studies). *Yes Energy*. https://blog.yesenergy.com/yeblog/the-utility-of-ai/ml-for-complex-energy-systems

NASA. (n.d.). Artificial intelligence (AI). *NASA Earthdata*. https://www.earthdata.nasa.gov/technology/artificial-intelligence-ai

Harvard Business Review. (2020, October 30). A practical guide to building ethical AI. *Harvard Business Review*. https://hbr.org/2020/10/a-practical-guide-to-building-ethical-ai

McKinsey & Company. (2023). Applying artificial intelligence for social good. *McKinsey & Company*. https://www.mckinsey.com/featured-insights/artificial-intelligence/applying-artificial-intelligence-for-social-good

AI.gov. (2023). Enhancing AI literacy for the United States of America. *AI.gov*. https://ai.gov/wp-content/uploads/2023/12/Recommendations_Enhancing-Artificial-Intelligence-Literacy-for-the-United-States-of-America.pdf

International Monetary Fund. (2024, January 14). AI will transform the global economy. Let's make sure it benefits humanity. *IMF*. https://www.imf.org/en/Blogs/Articles/2024/01/14/ai-will-transform-the-global-economy-lets-make-sure-it-benefits-humanity

Klawe, M. (2020, July 16). Why diversity in AI is so important. *Forbes*. https://www.forbes.com/sites/mariaklawe/2020/07/16/why-diversity-in-ai-is-so-important/